Theorising Cyber (In)Security

This book argues that cybersecurity's informational ontology offers empirical challenges, and introduces a new interdisciplinary theoretical and conceptual framework of 'entropic security'.

Cyber-attacks have been growing exponentially in number and sophistication; ranging from those conducted by non-state actors to state-backed cyber-attacks. Accordingly, cybersecurity now constitutes an integral part of public, private, and academic discourses on contemporary (in)security. Yet, because its emergence as a novel security field occurred after many long-established frameworks had already been developed, cybersecurity has been repeatedly scrutinised for its compatibility with conventional security theories, concepts, and understandings, particularly with that of military security. This book, however, argues that cybersecurity differs profoundly from many other security sectors because of the ontological nature of 'information' that sits at the heart of this field. Through this new framework, the book investigates three key empirical challenges in cybersecurity that are co-produced by its informational ontology: (1) the disordered nature of cybersecurity and its tendency towards increasing insecurity as a manifestation of the intrinsic uncertainties in information systems; (2) the unpredictable and unintended consequences resulting from autonomous cyber-attacks that challenge human control of cybersecurity environments; and (3) the persistent harms engendered by 'mundane' cyber threats that do not fit within conventional understandings of existentiality in security theories. Through a detailed analysis of cybersecurity discourses and practices in the USA (2003-present), the book goes on to show how these complex cybersecurity challenges are better analysed and theorised through the new information-theoretic notion of 'entropic security'.

This book will be of much interest to students of cyber-security, critical security studies, science and technology studies and International Relations in general.

Noran Shafik Fouad is a Senior Lecturer in Digital Politics at Manchester Metropolitan University, UK, and has a PhD in International Relations from the University of Sussex, UK.

Theorising Cyber (In)Security

Information, Materiality, and Entropic Security

Noran Shafik Fouad

Routledge
Taylor & Francis Group

LONDON AND NEW YORK

First published 2025
by Routledge
4 Park Square, Milton Park, Abingdon, Oxon OX14 4RN

and by Routledge
605 Third Avenue, New York, NY 10158

Routledge is an imprint of the Taylor & Francis Group, an informa business

British Library Cataloguing-in-Publication Data
A catalogue record for this book is available from the British Library

ISBN: 978-1-032-59308-1 (hbk)
ISBN: 978-1-032-59309-8 (pbk)
ISBN: 978-1-003-45411-3 (ebk)

DOI: 10.4324/9781003454113

Typeset in Times New Roman
by SPi Technologies India Pvt Ltd (Straive)

Contents

Acknowledgments

This book is based on my PhD project; a wonderful journey for which I am grateful to many people. First and foremost are my amazing PhD supervisors: Professor Stefan Elbe and Dr Stefanie Ortmann. I thank them for their immense support, encouragement, patience, and guidance in all stages of my PhD, and even after. In addition to their insightful feedback and stimulating discussions, they gave me confidence in my research abilities and in my capacity to become a good academic, and for that I cannot thank them enough. It was in Stefan's office that I first saw the *Routledge Handbook of Philosophy of Information*, which completely changed the trajectory of my research project. And it was Stefan's encouragement that gave me the courage to frame my project around entropy and to jump into exciting yet intimidating intellectual territories. Stefan remains a mentor to me and a role model for the academic I aspire to be. I extend my gratitude to Dr Shane Brighton, who co-supervised my thesis in the first year. Shane's and Stefanie's challenging feedback prompted me to develop my thinking about the project in such a way that completely transformed my initial research proposal.

I was also lucky to have a very rich discussion about the project with two fantastic security scholars as part of my PhD viva: Professor Anna Stavrianakis and Dr Tim Stevens. Their questions and comments influenced my thinking about the project since the viva and until the writing of this book. Although I passed without corrections, I thank Tim for the extra time he took to share extensive comments with me after the viva so I can address them in future publications. Tim's writings have inspired me in many ways in thinking about cybersecurity through a philosophical/theoretical lens, and I am grateful for the opportunity to have had him thoroughly engage with my work.

This project also benefited from feedback, discussions, and conversations in many academic contexts. I especially want to thank the participants of the GLOBE International Relations Winter School at the University of Lausanne (2019); the participants of the Sussex-Copenhagen PhD Workshop at the University of Copenhagen (2018); the participants of 17th and 18th European Conference on Cyber Warfare and Security (2018, 2019); and the participants of the BISA Postgraduate and Early Career Researchers Conference (2019).

This book also draws upon **Fouad, N. (2022). The non-anthropocentric informational agents: Codes, software, and the logic of emergence in cybersecurity. *Review of International Studies*, 48(4), 766-785. © The Author(s), 2021. Published by Cambridge University Press on behalf of the British International Studies Association.** I am grateful to the anonymous reviewers of this article and their constructive comments that helped me develop some of the arguments that appear in this book. I extend this gratitude to the three anonymous reviewers who engaged with my book manuscript for all their comments, questions, and suggestions.

I conducted this project at the University of Sussex in the Department of International Relations; a wonderful place and an incredibly stimulating and supportive academic community that I am lucky I was part of. This is where I developed a fascination with critical approaches to security and IR and learnt to view critique as integral to the study of social and political phenomenon rather than an alternative perspective. I also gratefully acknowledge the funding I received from the University to conduct this research through the Chancellor International Research Scholarship (CIRS), which made this journey possible.

The PhD thesis and the book could not have come to life if it was not for the support and love I have always received from fantastic friends and my family who believed in me way more than I could ever believe in myself. I will not mention names, but you will recognise yourself if you remember asking me at least once every other week: *'how is the book coming?'*

Acronyms and Abbreviations

AI	Artificial Intelligence
ANT	Actor-Network Theory
ARPA	Advanced Research Projects Agency
BRL	Ballistic Research Laboratory
CNA	Computer Network Attack
CND	Computer Network Defence
CNE	Computer Network Exploitation
CNIs	Critical National Infrastructures
CNAs	Computer Network Attacks
CND	Computer Network Defence
CNE	Computer Network Exploitation
CNOs	Computer Network Operations
CERT	Computer Emergency Response Team
CSS	Critical Security Studies
DARPA	Defence Advanced Research Projects Agency, the USA
DDOS	Distributed Denial of Service attacks
DHS	Department of Homeland Security, the USA
DoD	Department of Defence, the USA
EDVAC	Electronic Discrete Variable Automatic Computer
ENIAC	Electronic Numerical Integrator and Computer
IBM	International Business Machines Corporation
ICSs	Industrial Control Systems
ICTs	Information and Communication Technologies
IDSs	Intrusion Detection Systems
IoTs	Internet of Things
IP	Intellectual Property
IPSs	Intrusion Prevention Systems
IR	International Relations
MIT	Massachusetts Institute of Technology
NHS	National Health Service
NSA	National Security Agency, the USA
NSF	National Science Foundation
ONR	Office of Naval Research

PPPs	Public-Private Partnerships
SAGE	Semi-Automatic Ground Environment
STS	Science and Technology Studies
UK	United Kingdom
USA	United States of America
WWII	Second World War
WWW	World Wide Web

1 Introduction

We live in a world with an ever-growing dependence on information and communication technologies (ICTs) and with "information everywhere" (Tarasewich & Warkentin, 1999). With dependency comes a large set of vulnerabilities and risks of serious consequences if the information systems we depend on are disrupted in anyway. Cyber threats are now major sources of such disruptions, with an increasing number of cyber incidents affecting individuals, public and private organisations, and governments. These threats are proliferating across sectors, leading to privacy breaches, financial losses, intellectual property (IP) theft, and serious operational disruptions. Even if individuals or entities manage to apply all the necessary measures to secure their data and/or systems, there is no guarantee they would not be subjected to cyber threats, either ones that directly affect them or others that affect the entities that they depend on. For example, most recently, a pathology laboratory that processes blood tests for the National Health Service (NHS) in the United Kingdom (UK), named Synnovis, was subject to a cyber-attack that resulted in publishing sensitive patients' data online and subsequently postponing a significant number of critical surgeries in NHS hospitals (*The Guardian*, 2024). It was reported that a senior manager in the NHS described this attack as "everyone's worst nightmare" (Campbell, 2024). This is not an isolated incident, however; it is one among many affecting all sectors, including finance, manufacturing, education, etc. As a result of such incidents, cybersecurity has become integral to the study of contemporary security and insecurity in International Relations (IR).

For many years, cybersecurity has mostly been approached in IR through the lenses of traditional security theories and concepts. Because its emergence as a novel security field occurred after many of the long-established theoretical and methodological frameworks in IR and Security Studies were already developed, cybersecurity has always faced the challenge of "fitting in." It has therefore been repeatedly scrutinised for its compatibility with conventional security logics and understandings, particularly with that of military security. Many literatures study cybersecurity by employing the conceptual frameworks of war, terrorism, and deterrence, and by assessing the gravity of the cyber threat using attack-based conceptualisations and traditional forms of violence as benchmarks (see Carr, 2012; Farwell & Rohozinski, 2011; Gartzke, 2013;

DOI: 10.4324/9781003454113-1

Henschke, 2021; Kaplan, 2017; Lindsay, 2013; McGraw, 2013; Nye, 2017; Rid, 2012). Imposing these militaristic frameworks onto the field of cybersecurity has revealed multiple tensions, however. In particular, the inherent complexity and technicality of cybersecurity as field of research and policy analysis have proven challenging to theorise and problematic to fit within conventional approaches to security.

Today, the majority of cybersecurity literatures are policy-oriented in nature and tend to be conceptually under-theorised. Due to the empirical nature of the field, there is not enough space for research questions that do not have immediate policy relevance or ones that do not attend to current events (Cristiano et al., 2024). It remains true that there is a growing body of theoretical contributions that generate a conceptually far more sophisticated approach to the study of cybersecurity when compared to the policy-oriented literature. Yet, such theorisation is often done by primarily resorting to theories and concepts in existing IR, Security Studies, and Science and Technology Studies (STS) literatures that have been long applied to other (conventional) security sectors and issues.

This book, however, argues that cybersecurity differs profoundly from many other security sectors because of the peculiar ontological nature of "information" that sits at the heart of this burgeoning field. Building upon recent scholarship on the *philosophy of information, information sciences*, and *software studies*, the book takes existing theoretical contributions a step further by introducing a novel, interdisciplinary theoretical and conceptual framework to the study of cybersecurity that captures its technicality, complexities, and peculiarities. As will be demonstrated throughout the book, attending to 'information' as the core subject matter and referent object (i.e., object of protection) of cybersecurity adds important insights to its theorisation and to the understanding of the material conditions that influence its socio-political construction. Accordingly, the book asks a key question: *what does security look like when the peculiar informational ontology of cybersecurity is acknowledged, and when information per se is even theorised as a generative force of its own (in)security?* In answering this question, a novel information-theoretic framework to the study of cybersecurity is advanced by theorising cybersecurity as *entropic security*, in which security is practised through the logics of *negentropy*, *emergence*, and *noise* – concepts that will be explained and fleshed out throughout this book.

Entropy is a concept that first originated in thermodynamics, but later moved to information theory and other academic fields, including economics, geography, and social theory. Although it is defined differently in different sciences, entropy generally denotes uncertainty, disorder, noise, or disorganisation (Li and Du, 2017). Building upon the various conceptualisations of entropy in information sciences, and through an analysis of cybersecurity discourses and practices in the United States of America (2003–present), the book develops the notion of *entropic security* to theoretically capture three empirical challenges in cybersecurity that are co-produced by its informational ontology: (1) the disordered nature of cybersecurity, its intrinsic uncertainties, and its tendency towards increasing insecurity; (2) the unpredictable and unintended consequences resulting from autonomous cyber-attacks that challenge human control of cybersecurity environments; and (3) the persistent harms engendered by 'mundane' cyber threats that do not fit within conventional understandings of existentiality in security theories. This is a theoretical framework that can be used to study cybersecurity in a wide variety of contexts, or even applied to other security issues. By initiating this interdisciplinary enquiry, the book speaks to scholars and post-graduate students in the fields of cybersecurity, Critical Security Studies (CSS), STS, the philosophy of information, software studies, and IR theory.

This introductory chapter starts by exploring the conceptual and policy challenges of cybersecurity and problematising its under-theorisation. It then moves to an analysis of the key limitations of security theories when applied to cybersecurity and how the book addresses those limitations as a basis for the alternative theorisation of cybersecurity it develops. Accordingly, the chapter introduces the concept of the "infosphere" to conceptualise cybersecurity as an ontologically differentiated field that is constituted, conceptualised, experienced, and managed through "information," and explains the main arguments of the book and the methodology through which it will be tested. The chapter further discusses key ontological and epistemological consideration in theorising cybersecurity in general and within the framework advanced by this book in particular. The chapter ends, finally, with an explanation of the book outline and chapters' division.

Problematising cyber (in)security

Ever since the "Morris Worm" first hit the earliest manifestation of the internet, the ARPANET, in 1988, hostile cyber operations have been growing exponentially in both number and sophistication; ranging from those conducted by non-state actors to state-backed attacks.[1] Concurrently, the range of "insecure" objects has also widened considerably to include not only governments, but also individuals, businesses, and electoral processes. Operations have been targeting the multiple layers of what is referred to as "cyberspace." These include physical systems (computers, cables, routers, and all hardware), virtual spaces (programs, codes, protocols, and all software), the cognitive domain (data, ideas, and meanings on digital systems), as well as the human users of information. These hostile cyber operations take a wide variety of forms. For example, a very common and widely known form is phishing campaigns, also known as social engineering, that trick targets into submitting personal or financial data or download malicious files to their systems. Operations known as Distributed Denial of Service attacks (DDOS) can flood a certain computer server with requests and stop it from providing services to its intended users. Cyber espionage is yet another example, in which information is obtained without the target's consent. Attacks using ransomware, a type of malicious software (malware), can block the target's ability to access their data and threaten to either delete it or release it on the internet unless a ransom is paid. This list could be extended much further.

During the past few years, several high-profile or allegedly state-backed operations were repeatedly reported by the media. These include the breach of the Democratic National Committee (DNC) in the United States of America (USA) in 2015 and 2016, the WannaCry ransomware attack which affected more than 200,000 computers in 150 countries, and the NotPetya, widely considered the costliest cyber incident in history with an estimated loss of 10 billion dollars (Kaspersky, 2018). Most recently, the USA and the UK, with support from Western allies, accused "Chinese-state affiliated" organisations and individuals of launching extensive cyber operations for many years against political targets in Washington and Westminster, including government officials, parliamentarians, and electoral commissions (Fisher et al., 2024).

However, the scope of the cybersecurity challenge is much wider still. Cybersecurity is as much about the less-than high-profile operations, as it is about the highly publicised ones. Such lower-level incidents take place on daily basis and may not even be discovered or reported by the targets, and hence not covered by the media. Based on a recent IBM report, for example, the average cost of data breach is now at an all-time high at $4.45 million, with entities in the USA experiencing the high average cost at $9.48 million (IBM, 2023). It is also reported that in the first half of the year 2023, ransomware attacks increased by 50% year-on-year, due to a growing use of Ransomware-as-a-Service (RaaS). RaaS is a business model in which groups with no technical skills to develop ransomware can buy kits from RaaS operators to launch such attacks, with prices as low as 40$ (World Economic Forum, 2024).

As a consequence, cybersecurity has risen in prominence on the agenda of governments around the world, and it now constitutes an integral part of public, private, and academic discourses on contemporary (in)security. As a field for contemporary security studies, moreover, cybersecurity is marked by divergent approaches and differing opinions. Academic debate is significantly divided, for example, on the nature and extent of the security problem. Whilst "cyber sceptics" criticise what they believe to be "cyber hype" and question the damaging effects of cyber-attacks to date (Gartzke, 2013; Lee & Rid, 2014; Lindsay, 2013; Rid, 2013), there are also those who believe that "cyberspace" has revolutionised modern conflict and should be dealt with as a whole new domain of warfare (Carr, 2012; Clarke & Knake, 2010; Junio, 2013; McGraw, 2013). Furthermore, inherent multidisciplinarity and complexity have led to significant contention – both on the academic and policy levels – about how to define cybersecurity, the relevant importance of referent objects it comprises, and the nature of threats it should be defended against.

Added to these conceptual debates are policy-related ones, stemming in part from increasing cyber dependencies and the massive development of ICTs. Those developments have widened the scope of potential attacks (Agrafiotis et al., 2018); lowered their costs and entry barriers to malicious threat actors

(Egloff, 2022); and ultimately making cyber defence more challenging (AlDaajeh & Alrabaee, 2024). Complex policy challenges also result from the need for information-sharing (Brilingaitė et al., 2022); public-private partnerships (Carr, 2016); the challenges of attribution on one side (Tsagourias & Farrell, 2020) and deterrence on the other (Nye, 2017); and the absence of acceptable international norms for states' behaviour in cyberspace (Kulikova, 2021). Furthermore, governments' intrusions in rivalries' networks, and their involvement in the black markets of vulnerabilities and zero-day exploits, have produced new threat discourses and questions of response.[2] These practices have now been normalised as part of state "defence" and are therefore not adequately questioned by academics and security experts, despite their role in undermining human security and trust among states (Buchanan, 2016; Burton & Lain, 2020; Dunn Cavelty, 2016; Egloff & Shires, 2021; Healey, 2019; Herzog & Schmid, 2016). On the other hand, the market-oriented views of the private sector, which tend to prioritise functionality over security, have also led to a culture of acceptance of software insecurity (Chong, 2016). This normalisation of private and public insecurity has created a situation in which patching (fixing) vulnerabilities (i.e., exploitable coding errors) on a regular basis is not yet an adopted behaviour by all private sector organisations, especially the developers and operators of the Industrial Control Systems (ICS) that run critical national infrastructures (CNIs) (Lee, 2016).

Security theories and cybersecurity

When cybersecurity started to grow as a significant issue in public, private, and academic discourses on contemporary (in)security, many literatures in Security Studies attempted to theorise it using pre-existing security theories and concepts. Early attempts to theorise cybersecurity employed discursive methodologies to explore how threat representations are different from other security sectors. A prominent example in this regard is literature that build upon the Copenhagen School's securitisation theory to theorise cybersecurity as a new and peculiar security sector. Securitisation theory defines security as a process through which a securitising actor presents an issue as posing an *existential threat* to a referent object, requiring *extraordinary measures* to ensure the object's *survival*. Thus, for an issue to qualify as "security," it has to be put "above politics" or to be presented as 'a special kind of politics' and has to be accepted by security audience. This process can be studied by analysing security as a *speech act* and a discourse that comprises social and political construction of threats (Buzan et al., 1998). Although cybersecurity was not part of its original formulation, many studies applied the theory's framework to understand the process of cyber securitisation, particularly in the USA as a case study (Bendrath et al., 2007; Dunn Cavelty, 2008a, 2008b; Eriksson, 2001; Hansen & Nissenbaum, 2009). More recent contributions called for applying the securitisation theory to understand the complexities of cybersecurity in "the non-West" (Lacy & Prince, 2018), and extended the discussion on cyber securitisation to other contexts, such as Singapore (Kallender & Hughes, 2017), Japan (Aljunied, 2019), and Egypt (Hassib & Alnemr, 2021).

Related to this cyber securitisation literature is a wide range of studies that use the securitisation theory's discursive methodology – if not the theory as such – to explore how cybersecurity discourses, utterances, and threat representations are different from other sectors. They note that cybersecurity discourses operate in the absence of a minimum level of agreement on the nature of threats, and sometimes with no empirical evidence of attacks to justify them. That is why such discourses mostly rely on symbolisations, by drawing comparisons between cyber threats and other conventional ones, characterised by "stable threat conventions" (Emerson, 2016; Hare, 2009; Jarvis et al., 2016). Added to this is the biologisation of technology and the use of "viruses" and "worms" metaphors (Dunn Cavelty, 2013); the spatial analogies of cyberspace (Betz & Stevens, 2013); viewing cybersecurity as inherently ungovernable and anarchic (Barnard-Wills & Ashenden, 2012); and the use of fear-based analogies and hypothetical cyber-doom scenarios, such as cyber 9/11 or cyber Katerina (Lawson, 2013; Lawson, 2019). Lene Hansen and Helen Nissenbaum's theorisation of cybersecurity as a *distinct* security sector by demonstrating its unique 'security grammars' is one notable contribution in this regard (Hansen & Nissenbaum, 2009).[3]

However, the emphasis on the discursive construction of security by the securitisation theory and its applications on various security issues has been problematised by multiple studies, often classified as the "second-generation" (Stritzel & Chang, 2015, p. 550) or the sociological model of securitisation (Balzacq, 2009). They criticise the theory for dismissing the extra-discursive and socio-political contextual influences on processes and practices of securitisation. Hence, they suggest different ways to incorporate those contextual influences by analysing the macro and micro environments of securitisation (Balzacq, 2011; Wilkinson, 2011); actor-audience relationships (Balzacq, 2005; Balzacq et al., 2015); forms of resistance to security frames (Stritzel & Chang, 2015); and the multiplicity of securitising audiences (Salter, 2008). In the same vein, the application of the theory to the study of cybersecurity has been criticised for its limited conceptualisation of the securitising actors whose discourses are relevant to the construction of security. This is seen in the cyber securitisation literature's preoccupation with official and government's discourses; thus, overlooking the role of non-state, private actors in producing and managing cybersecurity discourses (Dunn Cavelty, 2016).

Importantly, there are several material realities in cybersecurity in relation to the nature of computer disruptions, their effects, and knowledge about them in technical communities that cannot be understood as part of discursive constructions alone (Dunn Cavelty, 2019). This includes, for example, the exponential rise in the number of cyber operations through self-replicating malware that enjoy a considerable level of autonomy in execution, and that spread beyond their intended target and cause unintended consequences. In fact, the very idea of computer viruses and worms – that constitute "the cyber weapon" – is an exemplar of how information systems are capable of deviating from the human intentionality embedded in their design, since they were not initially designed to be used maliciously. Cyber incidents caused by malware are, in turn, major challenges to ideas of control upon which computing technologies were based. As argued by Thomas Rid, cyber incidents are the dystopia of the promises of "cybernated" economies, cyborgs, and cyberspace as a new parallel frontier to reality. Now, control over machines can be taken from humans, systems can be attacked and controlled distantly, and several damages can result in the form of data loss, abuse, denial of services, or even machine damaging (Rid, 2016).

Therefore, more recent theoretical contributions in theorising cybersecurity have shifted towards a critical engagement with the politics and *materialities* of cybersecurity, beyond rhetoric and discursive representations (Stevens, 2015). They approach cybersecurity as an "assemblage," or a complex configuration of actors (Collier, 2018). Cybersecurity assemblages, they argue, operate through various material, political, and social alliances that make malware successful and that prioritise particular cybersecurity incidents and not others (Egloff & Dunn Cavelty, 2021; Jacobsen, 2020; Stevens, 2020). Other newly emerging literatures theorise cybersecurity by focussing on 'non-human' actors, such as codes, software, or malware, and conceptualising them as political agents, capable of co-constructing the "space" in cyberspace and co-shaping the politics of cybersecurity (Balzacq & Dunn Cavelty, 2016; Dwyer, 2021; Fouad, 2022; Liebetrau & Christensen, 2021).

Against this background, and instead of using pre-existing concepts and theories in IR and Security Studies to theorise cybersecurity, this book takes the aforementioned theoretical contributions a step further to develop a new theoretical and conceptual framework to study cybersecurity in a way that attends to its multidisciplinarity, technicality, and complexity. It does so by developing an information-theoretic conceptual and theoretical framework that analyses cybersecurity as an ontologically informational field. Contending that cybersecurity is inherently informational means that it is essentially constituted, conceptualised, experienced, and managed through information as its core subject matter and referent object. Investigating this informational ontology of cybersecurity is, in many ways, ultimately a study of *materiality*.

The book studies materiality of cybersecurity in two main ways. Firstly, it does so by acknowledging the informational essence of the "cyber," and its constitutive technologies and sciences, beyond speech acts and linguistic utterances. As argued by Bennett, studying materiality means acknowledging that non-humans are real agents or actors rather than social constructs or mere instruments (Bennett, 2010). Thus, the book considers the role of information and its properties as such in shaping, enabling, and/or limiting the construction processes of cybersecurity discourses and practices. Here, the book defines information by its syntactic (signs, signals, bits, etc.), semantic (meanings conveyed through those bits), and pragmatic elements (signifying the relationship between meanings and the receiver's knowledge) (Deacon, 2010). This definition and the various other ways of defining information will be further explored in the next chapters. Secondly, the book approaches information as a *peculiar* referent object of cybersecurity. This peculiarity is examined in light of three key *ontological* properties of information: (1) the intrinsic indeterminacies surrounding the operation of information systems; (2) the complexities and non-linearities, of syntactic elements of information (i.e., codes/software) and the contingencies they produce; and (3) the simultaneous physicality and non-physicality of information.

These peculiar properties of information, the book argues, generate security logics that are quite different from conventional accounts of security in IR and Security Studies. The logics of existentiality, exceptionality, and emergency that are used in some theoretical literature are not readily applicable to the field of cybersecurity, as they undermine the inherent informational nature of cybersecurity and how information co-constructs different logics of (in) security. Considering the omnipresent prominence of cybersecurity in policy and academic debates, as mentioned above, a reconsideration of the criteria of "securityness" is essential for any attempt to theorise this field. This is an aspect that Dunn Cavelty, a key scholar in the cyber securitisation literature, has acknowledged in a later work (Dunn Cavelty, 2020).

To capture the distinct security logics co-produced by the peculiarities of information in cybersecurity, the book develops the information-theoretic notion of *entropic security*. Entropy is a concept that first originated in thermodynamics, but later moved to information theory and several other academic fields, including economics, geography, and social theory. In information theory and cybernetics, entropy is mostly defined as uncertainty and disorder, randomness and non-linearity, or disruption in communication channels (Li & Du, 2017). The book uses these three definitions to advance an understanding of cybersecurity as entropic security, practised through the logics of *negentropy*, *emergence*, and *noise* – all of which will be explained and fleshed out in the next chapters.[4]

First, the book examines the intrinsic uncertainties and tendency towards disorder in the operation of information systems, and how this ontological property influences the *meaning* and *essence* of "security" in cybersecurity. On that basis, cyber defence practices are reconceptualised as "anti-entropic practices" directed against the entropic force of increasing disorder and insecurity; i.e., aiming at *negentropy* (negative entropy). Second, the book studies the entropic nature of cybersecurity by investigating the randomness, contingencies, and non-linearity generated by the operation of codes/software, as a form of syntactic information. A particular focus is to the role of malware in co-producing enmity and the subjects/objects of cybersecurity through what the book calls the logic of *emergence*. Third, entropy as disruption in communication channels is used analogically in highlighting the significance of *mundane* cybersecurity as opposed to existential threats that mark other security fields. In this respect, the book analyses the (non-)physicality of information and its impact in co-constructing cyber threats through the logic of *noise* that can invoke urgency without existentiality. Negentropy, emergence, and noise are all manifestations of the notion of entropy that challenge conventional understandings of security and that are more closely connected to the sciences and technologies constitutive of the "cyber."

From a discursively constructed sector to an "infosphere"

Sectoralisation, or studying security through the framework of sectors, has been integral to the evolution of security studies away from conventional theories. As Buzan and Little argue: "In IR, sectoral analysis refers to the practice of approaching the international system in terms of the type of activities, units, interactions, and structures within it" (Buzan & Little, 1998, p. 72). For example, in moving away from the military sector in studying security, which was the primary focus of classical theories of IR, the Copenhagen School's securitisation theory added other sectors to security theorisation: the political, economic, societal, and environmental sectors. In Albert and Buzan's work on sectoralisation, they asked an important question on whether security sectors are mere analytical lenses which overlap, or ontological realms that have autonomous existence (Albert & Buzan, 2011). The two scholars did not give a definitive answer, but in their discussion, they limited the criteria of sector differentiation to human actors, their perceptions, and their discourses. For example, they argued that sectors like the political or the economic could be classified as "ontologically real" because they have a specific basal code; e.g., a logic for what constitutes having or not having power in a political system. According to them, the same argument cannot be made about the environmental or societal sector, given the absence of this communicational basal code. In short, they approached sectoralisation as a fundamentally empirical question, whose answer may vary according to place, time, and actors' discourses.

Although sectoralisation played an important role in widening the agenda of security studies beyond military security, it remains problematic given its intrinsic link to human-centric discursive constructions. On the contrary, and without renouncing the inherent connections between cybersecurity and other security sectors, this book presents an argument that cybersecurity is ontologically differentiated by its fundamentally *informational* nature – an assumption that will be explored in the next chapters. Focussing on this informational ontology, rather than human speech acts, allows for theorising the peculiarities of cybersecurity and investigating its ontological makeup in ways that are not currently achieved by the existing security literature. This theoretical move from the "cyber" to the "informational" is also necessary because it provides a deeper account for the multidisciplinarity of cybersecurity and its inherent links with information sciences.

Hence, the book presents a novel interdisciplinary exploration of cybersecurity by bringing the philosophy of information into IR and Security Studies. The philosophy of information, as a newly emerging field of research, interrogates the concept of information and provides important philosophical insights about its nature, principles, and dynamics (Adriaans & van Benthem, 2008;

Floridi, 2010, 2013, 2016). This field evolved with the massive development of ICTs, and particularly computing and internetworking technologies. These developments brought information to the centre of philosophy as one significant force in the functioning of the world. Through the philosophy of information, one can study the structure of information and its representation in both machines and humans. Though "philosophy of information" can refer to the philosophical study of information sciences – in the same way we can study the philosophy of humanities, for instance – this field also aims at presenting "information as a major category of thought within philosophy itself" (Adriaans & van Benthem, 2008, pp. 3–4). That is, the philosophy of information intends to use an "information-oriented stance" in approaching epistemological and ontological questions in the world.

Similarly, the book adopts an information-oriented stance in studying cybersecurity by establishing a dialogue between the philosophy of information on one side and IR and Security Studies on the other. Here, cybersecurity is presented as an "infosphere" rather than a discursively constructed sector. The notion of the "infosphere" is drawn from the work of Luciano Floridi (2009, 2010, 2013, 2014), a professor of philosophy and ethics of information and one of the prominent contributors to this multidisciplinary field.[5] Floridi justifies the need for a philosophy of information by looking differently at the role of ICTs in our world. He assumes that ICTs are not simply "enhancing" human life, but rather "re-ontologising" it. By re-ontologisation Floridi refers to the fundamental transformations to reality and to humans as a result of the information revolution, that he captures through the concept of the infosphere (Floridi, 2010). An infosphere, as described by Floridi, is an informational environment combining several informational entities that interact with one another in both online and offline spaces. It combines all those entities' properties, processes, and relations (Floridi, 2014). Although Floridi was referring to existence in its totality when he introduced the idea of the infosphere, this concept can also contribute to developing an informational approach to the study of cybersecurity. That is because Floridi's infosphere resembles cybersecurity in a number of ways.

Firstly, the infosphere is "hyperhistorical." This marks a transition from pre-history, when no ICTs existed; to history, in which progress and welfare is *related* to ICTs; and then finally to hyperhistory, when progress is *dependent* on ICTs. In this stage, ICTs are not just important, but a pre-requisite for economic and societal development. It is also characterised by an exponential rise in the amount of data that needs to be stored and processed, which Floridi called the flood of "zettabytes"; commonly referred to as 'big data' (Floridi, 2014). Likewise, it can be argued that as a hyperhistorical infosphere, cybersecurity threats are driven by the increasing dependence on information technologies and big data. In this respect, cybersecurity is different from security sectors in which more development may bring more security, be it economic, military, or political development. As Floridi puts it, "Only a society that lives hyperhistorically can be threatened informationally, by a cyber attack. Only those who live by the digit may die by the digit" (Floridi, 2014, p. 4).

Secondly, Floridi's infosphere witnesses what he calls 'third-order technologies' and the erosion between online and offline existence. It signifies the state of technology as both a user and prompter of innovation, such as the case of the internet of things (IoTs), in which humans are kept outside the loop of communications. Devices connect to one another, exchange protocols, send and receive data, update their files, all possibly without the intervention of the human beneficiary. Eventually, "being out of the loop could mean being out of control" (Floridi, 2014, p. 39). Thus, instead of being merely tools that facilitate human life, ICTs are becoming forces of their own that shape reality. Humans in turn become informational organisms, or *inforgs*, that interact with a variety of other non-human informational agents. Many aspects of cybersecurity are fundamentally ungovernable and/or uncontrollable by humans. That is why cybersecurity, understood as an infosphere, should be approached as a field of contingencies that ultimately escape the span of absolute human control. One possible political implication of this is shifting the essence of politics and security from managing people's lives towards managing the "life cycle of information." This life cycle of information includes information occurrence, transmission, processing, and usage (Durante, 2017).

Finally, in an infosphere, power over data and ICTs does not reside in the state as the sole agent; it is distributed among a wide range of non-state actors, enabled and empowered by such technologies (Floridi, 2014). Hence, the analysis of the state as a political organisation is no longer the core of understanding politics, rather, it is the "organisation of relationships between agents of different types and natures" (Durante, 2017). The book extends this argument to cybersecurity as an infosphere that ultimately breaks the key dichotomies of subject/object, human/non-human, and public/private. As put by Floridi, "ICTs are not merely re-engineering but actually re-ontologizing our world" (Floridi, 2010, 10–11). There is a subsequent need then for studying the impact of this re-ontologisation on the study of security at large, and specifically on the study of ICTs security as such. Through its interdisciplinary informational framework and its information-theoretic notion of entropic security, the book thus attends to the peculiar ontology of cybersecurity by transcending the limits of discursive analysis that have characterised a big part of the scholarship on cybersecurity to date.

Design and methodology

To summarise, the central argument of this book is that the *informational* ontology of cybersecurity poses profound conceptual challenges to traditional theorisation of security and co-produces distinct logics of (in)security. The next chapters will therefore develop this information-theoretic analysis of cybersecurity by conceptualising it as *entropic security*, constructed through the three logics of *negentropy, emergence*, and *noise*. Crucially, this theoretical and conceptual framework builds on three main properties of information: the intrinsic indeterminacies of information systems; the complexities and non-linearities of codes/software; and the simultaneous physicality and non-physicality of information. Each of these properties will be linked to an empirical challenge in cybersecurity: (1) intrinsic uncertainties and tendencies towards increasing insecurity in cybersecurity; (2) unpredictable and unintended consequences of autonomous cyber-attacks; and (3) persistent harms resulting from mundane cyber threats.

The book mobilises interdisciplinary literatures in the philosophy of information and information sciences – together with the closely related field of software studies which focusses on the social, cultural, and political impact of software (Fuller, 2008). It also draws upon this literature to introduce three specific security logics as manifestations of the notion of entropy – negentropy, emergence, and noise – that make up this field of cybersecurity. In addition, the book also draws upon literatures on new materialism and CSS to problematise the question of materiality in the study of cybersecurity.

This interdisciplinary theorisation will be illustrated in relation to cybersecurity discourses and practices in the USA since 2003, when the country's first cybersecurity strategy was announced (The White House, 2003). Multiple cybersecurity policy documents issued by the government are analysed, in addition to congressional hearings, in which several non-state actors testify, including members of the private sector, security experts, academics, and think tanks. Widening the scope of analysis to include non-state actors allows for an understanding of mundane cybersecurity beyond the existential, the exceptional, and the high-profile cyber incidents that are more evident in the military and intelligence discourses.

The principal official policy documents analysed include those issued by the White House: *Cybersecurity Strategy* (2003, 2023), *Cyberspace Policy Review* (2009), *Comprehensive National Cybersecurity Initiative* (2010), *International Strategy for Cyberspace* (2011), *Executive Order: Improving Critical Infrastructure Cybersecurity* (2013), and the *Presidential Policy Directive: Critical Infrastructure Security and Resilience* (2013). Added to this are documents issued by the Department of Defence (DoD): *National Military Strategy for Cyberspace Operations* (2006), *DoD strategy for Operating in Cyberspace* (2011), and *DoD Cyber Strategy* (2015, 2023). Finally, other documents include the *Blueprint for a Secure Cyber Future* (2011) and the *DHS Cyber Strategy (2023)* issued by the Department of Homeland Security (DHS). These are all the official strategy documents dealing strictly with cybersecurity and can be retrieved via these entities' official websites. Given the vast amount of congressional hearings that deal with cybersecurity-related issues, the book only focusses on those that include "cybersecurity" or "cyber" in their title in two committees: the Committee on Homeland security in the House of Representatives, and the Committee on Homeland Security and Governmental Affairs in the Senate. The choice of these two committees is based on their direct link to cybersecurity policy-making process and their general security nature, rather than being sector-specific, which matches the scope of this book.

Together with the theoretical literature on information, the qualitative analysis of the empirical data represents an understanding of discourse and materiality as essentially intertwined. As Karen Barad argues in her theory of agential realism, discursive practices are not exclusive to humans (Barad, 2003, 2007). Discourse is not synonymous with speech acts, rather, it is the force that enables/conditions those acts. Thus, if cybersecurity is to be studied as an infosphere in a non-anthropocentric approach, the *intra-action* between materiality and discourses has to be considered. That is, as Barad puts it, discourses are material and materiality is also discursive (Barad, 2003, 2007). More on the relationship between discourse and materiality, as well as their methodological implementation in the book, will be discussed further in Chapter 3.

Ontological and epistemological considerations

Security is "an essentially contested concept" (Buzan, 1991). This contested nature of security has been in itself relatively contested by scholars who argue that contestation is not unique to security but is a characteristic of all social order concepts (McSweeney, 1999), and that security is better designated as "confused or inadequately explicated" than contested (Baldwin, 1997). Nevertheless, assuming that security is essentially contested does not just refer to the challenge of defining it, which is certainly true to many other concepts, but importantly, it indicates the difficulty of doing so neutrally; i.e., without activating multiple assumptions about world politics and conflicting theoretical and normative positions (Smith, 2005). This raises questions on how security should be theorised and whether it can have a specific "meaning" or "logic."

CSS as a research agenda emphasises the importance of contextualising security. This requires an examination of how the meaning of security is always subject to negotiations and constant challenging, which may result in its eventual transformation (Stritzel, 2011). The assumption that security is transforming beyond any fixed logics that security theories may develop is proven by the historical genealogies of the term, which show that not only its meaning, but also its beneficiaries and implications, be them positive or negative, have been continuously changing over time and across different communities (McDonald, 2015). As argued by Ciută, there is a tension between assuming that security is the result of an intersubjective process, co-produced by multiple actors, and in the meantime specifying a fixed meaning for *that* security (Ciută, 2009). Contextualising security, therefore, means not imposing existing security logics, which are primarily tied to military security, to evaluate the 'securityness' of cyber discourses and practices. Instead, we should be investigating how cybersecurity can constitute its own security logics, that is both distinctive and transformative.

However, there remains some tension between analysing the ontology of cybersecurity and developing theoretical and conceptual frameworks to study its peculiarities on one side, and accepting the assumption that security is essentially contextual on the other. As argued above, one key contribution by

CSS, which this book speaks to, is the fundamental understanding that there is no one 'cybersecurity' everywhere and for everyone. Put differently, cybersecurity too is experienced differently across space, place, and time and by different actors (Cristiano et al., 2024; Dwyer et al., 2022). It follows that theorising cybersecurity as a generalisable phenomenon or attempting to develop "a theory" of cybersecurity would be essentially problematic. Accordingly, what this book advances here is not "a theory" of cybersecurity, but rather, an analysis of its informational ontology and how such ontology co-produces different security logics from that of conventional security. This is an attempt to contextualise logics of security in the study of cybersecurity as a relatively new field, without fixing such logics or generalising their applicability.

Similarly, entropic security as an information-theoretic framework is presented by the book as a useful approach to the study of security, especially cybersecurity, but one whose applicability should be tested and investigated in different political, social, and cultural contexts. Although the theorisation of entropic security is not case-specific, focussing the research design on an analysis of one in-depth case study is intended to increase the coherence of the results and facilitate the research process. The theoretical hypothesis of this book also requires an extensive analysis of cybersecurity discourses and practices of multiple actors over an extended period of time; and the ready availability of a vast amount of cybersecurity-related data in the USA makes it an ideal case study. Yet, despite the centrality of the case of the USA in the origination and evolution of cybersecurity debates to date, it remains one among many other important cases, and thus care must be taken in generalising from this one primarily Western-centric case study.

That is, the book proceeds with an assumption that information is a peculiar entity, but how this peculiarity is manifested empirically in the construction of cybersecurity is essentially context dependent. The theoretical framework advanced by this book can therefore lead to different conclusions when applied to different cases or contexts; something that can only be proven by further research. This can be particularly true in cases where the terminology of *information* is used to counter the arguably Western *cyber* terminology, such as in Russia and China. It would be also important to see if the books' arguments hold in cases that are traditionally seen as similar to the USA, such as the UK, and to be able to theoretically explain any possible deviances using the same informational framework.

Book outline

The remainder of this book proceeds in six chapters. *Chapter 2* provides an historical and a conceptual overview of cybersecurity as a field of study and policy analysis. It presents a brief historical account of the evolution of computing and internetworking technologies – which constitute what we commonly refer to as "cyberspace" – from the Second World War until the end of the Cold War. This account challenges the common and largely ahistorical approaches to studying cybersecurity by arguing that security has always been an integral part of the co-production of those technologies – as a context, generative force, and policy concern. It also shows how such information technologies evolved beyond the intentionality of their human inventors, which in turn influenced how their security was conceptualised in different historical stages. Moving from history to modern-day cybersecurity, the chapter analyses the current state of conceptualising cybersecurity for further clarification of the concept and distinguishing it from other related but analytically different terms, such as "internet security," "ICT security," "internet governance," among others. This is followed by an examination of the main policy challenges that define the current debates on cybersecurity and gaps in the literature that study those challenges.

Chapter 3 lays the foundation for theorising cybersecurity as an "infosphere" by establishing a conversation between (critical) security studies on one side and the newly emerging field of the philosophy of information on the other. As such, it introduces and substantiates three key theoretical assumptions/arguments. Firstly, cybersecurity is ontologically informational. Information lies at the heart of all the technologies, sciences, and practices that enable this field and its very existence. Secondly, information is a peculiar entity. It has an autonomous status that sets it apart from other non-human "things." Finally, it follows that cybersecurity should be theorised differently from other conventional security fields. This is because information and its peculiar properties co-produce distinct meanings and logic(s) for "security" in cybersecurity, which require new theoretical frameworks to understand. Building on these assumptions, the chapter challenges three conventional logics of security that have been used to study cybersecurity in theoretical literature: the logics of existentiality, exceptionality, and emergency. Here, the concept of "entropic security" is introduced, with three alternative logics of security that better capture the complexities and peculiarities of cybersecurity: negentropy, emergence, and noise. Each of these three logics and their relevance to the study of cybersecurity is advanced and substantiated in the next three chapters.

Chapter 4 builds upon the definition of entropy as uncertainty in information theory and as disorder in cybernetics to analyse the uncertainties of information systems and their tendency towards disorder and insecurity. Theoretically, the chapter analyses uncertainty as a property of information's existence that cannot be reduced to the empirical challenge of "not knowing." In practical terms, the chapter investigates the multiple sources of uncertainties in the operation of digital information systems and the kind of challenges they pose for cybersecurity policies, which the chapter frames as the "entropic space of cybersecurity." These include the uncertainties associated with vulnerability analysis, intrusion detection, attribution and damage analysis, and the lack of technical knowledge about such systems. Finally, the chapter examines how such properties co-produce a specific perception of defence in the infosphere that does not always need a pre-defined enemy or an attack; but rather is primarily exercised against the entropic force of increasing disorder and insecurity. Here, the logic of negentropy (i.e., negative entropy) is introduced as a revised understanding of cyber defence that frees it from conventional antagonistic, friend-enemy logics. Negentropy, as the essence of cybersecurity, is based on risk prioritisation and risk acceptance: accepting that inevitability of cyber threats, and thus directing most security measures and capabilities to the higher cyber risks.

Chapter 5 investigates the complexities of codes/software, as syntactic manifestations of information, their self-organising capacities, and their contingent properties to develop an understanding of cybersecurity as emergent security. Emergence is a key concept in complexity theory and the study of self-organising systems, and one important characteristic of entropy. It illuminates the inherent unpredictability of complex informational systems and the elements of novelty associated with their operations. The chapter introduces "emergence" as a non-linear logic that captures the contingencies of information and the uncertainties they engender in cybersecurity. It specifically shows how the logic of emergence and emergent security challenge the idea of human control in cybersecurity in two ways: by undermining the centrality of human intentionality as a basis for constructing enmity, and by acknowledging the role of codes/software in co-producing the subjects/objects of cybersecurity.

Chapter 6 analyses the simultaneous physicality and non-physicality as one ontological property of information. It shows how such property problematises the logic of existentiality and the question of survival in conventional security theories and their application on cybersecurity. In doing so, it investigates the different ways the physical and non-physical in digital information systems interact, change, and define one another. It further demonstrates the various security and privacy concerns that this complex (non-)physicality engenders, regarding the geolocation of data centres, data routing, undersea fiber-optic cables, and hardware/software manufacturing. Moving from general security implications, the chapter analyses how this (non-)physicality of information co-constructs the logic of noise that accentuates mundane cybersecurity threats in face of the existential. Viewed as disruption to communication channels in information theory, the concept of noise can be used to understand the complexity of cyber threats in a space between contingency and control. The same as noise is seen as a "normal" characteristic of information operation, but one that needs to be minimised and challenged, cyber threats are often viewed as a disruptive yet integral aspect of the everyday functioning of systems. Accordingly, the chapter shows how this property of information is capable of reducing existentiality to being just another discourse in cybersecurity and co-producing practices in which cyber threats are portrayed as urgent and immanent, albeit not existential. *Chapter 7* concludes the arguments presented in this book and highlights key contributions and prospects for further research.

Notes

1 For further information on the Morris Worm and its implications see (Orman, 2003).
2 Unknown vulnerabilities, or exploitable bugs in coding, are sometimes called "zero-day vulnerabilities" or shortly "zero-days." They are the vulnerablities that are unknown to the software vendors and for which no patch is available. Hence, the name "zero-day," which refers to the number of days the vulnerability was known to the target (Ablon & Bogart, 2017).
3 As argued by Hansen and Nissenbaum, the first cybersecurity grammar is *hypersecuritization*, through which cybersecurity discourses focus on disaster scenarios that have not taken place. The second is *everyday security practices*, by linking the scenarios of digital disasters to familiar experiences from everyday life, like credit card fraud, identity theft, etc. The third is *technification*, that creates political legitimacy for security experts by presenting cybersecurity as a domain that requires technical knowledge that the public do not have (Hansen & Nissenbaum, 2009).
4 Note that the notion of entropic security developed in this book is not related to "entropic security" in the field of cryptography. In cryptography, the notion of entropic security was introduced by Russell and Wang in 2002 to specifically refer to an encryption scheme that relies on the aggressor's uncertainty regarding the function of the transmitted message (Russell & Wang, 2002).
5 Floridi acknowledged that the concept goes back to the 1970s and has its roots in the concept of "biosphere," or the space on Earth in which life exists (Floridi, 2014).

References

Ablon, L., & Bogart, A. (2017). Zero Days, Thousands of Nights: The Life and Times of Zero-Day Vulnerabilities and Their Exploits. Rand Corporation.

Adriaans, P., & van Benthem, J. (Eds.). (2008). *Philosophy of Infomrmation* (Vol. 8). North Holland.

Agrafiotis, I., Nurse, J.R.C., Goldsmith, M., Creese, S., & Upton, D. (2018). A Taxonomy of Cyber-Harms: Defining the Impacts of Cyber-Attacks and Understanding How They Propagate. *Journal of Cybersecurity, 4*(1), tyy006.

Albert, M., & Buzan, B. (2011). Securitization, Sectors and Functional Differentiation. *Security Dialogue, 42*(4–5), 413–425.

AlDaajeh, S., & Alrabaee, S. (2024). Strategic Cybersecurity. *Computers & Security, 141*, 103845.

Aljunied, S.M.A. (2019). The Securitization of Cyberspace Governance in Singapore. *Asian Security, 0*(0), 1–20.

Baldwin, D.A. (1997). The Concept of Security. *Review of International Studies, 23*(1), 5–26.

Balzacq, T. (2005). The Three Faces of Securitization: Political Agency, Audience and Context. *European Journal of International Relations, 11*(2), 171–201.

Balzacq, T. (2009). Constructivism and Securitization Studies. In M.D. Cavelty & V. Mauer (Eds.), *The Routledge Handbook of Security Studies* (pp. 56–72). Routledge.

Balzacq, T. (2011). Enquiries into Methods: A New Framework for Securitization Analysis. In T. Balzacq (Ed.), *Securitization Theory: How Security Problems Emerge and Dissolve* (pp. 31–54). Routledge.

Balzacq, T., & Dunn Cavelty, M. (2016). A Theory of Actor-Network for Cyber-Security. *European Journal of International Security, 1*(02), 176–198.

Balzacq, T., Leonard, S., & Depauw, S. (2015). The Political Limits of Desecuritization: Security, Arms Trade, and the EU's Economic Targets. In T. Balzacq (Ed.), *Contesting Security: Strategies and Logics* (pp. 104–121). Routledge.

Barad, K. (2003). Posthumanist Performativity: Toward an Understanding of How Matter Comes to Matter. *Journal of Women in Culture and Society*, *28*(3), 801–831.

Barad, K. (2007). Meeting the Universe Halfway: Quantum Physics and the Entanglement of Matter and Meaning. Duke University Press.

Barnard-Wills, D., & Ashenden, D. (2012). Securing Virtual Space Cyber War, Cyber Terror, and Risk. *Space and Culture*, *15*(2), 110–123.

Bendrath, R., Eriksson, J., & Giacomello, G. (2007). From 'Cyberterrorism' to 'Cyberwar', Back and Forth: How the United States Securitized Cyberspace. In J. Eriksson & G. Giacomello (Eds.), *International Relations and Security in the Digital Age* (pp. 57–82).

Bennett, J. (2010). A Vitalist Stopover on the Way to a New Materialism. In D. Coole & S. Frost (Eds.), *New Materialisms: Ontology, Agency, and Politics* (pp. 47–69). Duke University Press.

Betz, D.J., & Stevens, T. (2013). Analogical Reasoning and Cyber Security. *Security Dialogue*, *44*(2), 147–164.

Brilingaitė, A., Bukauskas, L., Juozapavičius, A., & Kutka, E. (2022). Overcoming Information-Sharing Challenges in Cyber Defence Exercises. *Journal of Cybersecurity*, *8*(1).

Buchanan, B. (2016). The Cybersecurity Dilemma: Hacking, Trust and Fear Between Nations. C. Hurst & Co Publishers.

Burton, J., & Lain, C. (2020). Desecuritising Cybersecurity: Towards a Societal Approach. *Journal of Cyber Policy*, *5*(3), 449–470.

Buzan, B. (1991). People, States and Fear: An Agenda for International Security in the Post-Cold War Era. Harvester Wheatsheaf.

Buzan, B., & Little, R. (1998). International Systems In World History: Remaking the Study of International Relations. Oxford University Press.

Buzan, B., Wæver, O., & Wilde, J. de. (1998). *Security: A New Framework for Analysis*. Lynne Rienner Publishers.

Carr, J. (2012). *Inside Cyber Warfare* (2nd ed). O'Reilly.

Carr, M. (2016). Public–Private Partnerships in National Cyber-Security Strategies. *International Affairs*, *92*(1), 43–62.

Chong, J. (2016). Bad Code: Exploring Liability in Software Development. In R. Harrison, T. Herr, & R.J. Danzig (Eds.), *Cyber Insecurity: Navigating the Perils of the Next Information Age* (pp. 69–86). Rowman & Littlefield Publishers.

Ciută, F. (2009). Security and the Problem of Context: A Hermeneutical Critique of Securitisation Theory. *Review of International Studies*, *35*(02), 301.

Clarke, R.A., & Knake, R. (2010). Cyber War: The Next Threat to National Security and What to Do About It. HarperCollins.

Collier, J. (2018). Cyber Security Assemblages: A Framework for Understanding the Dynamic and Contested Nature of Security Provision. *Politics and Governance*, *6*(2), 13–21.

Cristiano, F., Kurowska, X., Stevens, T., Hurel, L.M., Fouad, N.S., Cavelty, M.D., Broeders, D., Liebetrau, T., & Shires, J. (2024). Cybersecurity and the Politics of Knowledge Production: Towards a Reflexive Practice. *Journal of Cyber Policy*, *0*(0), 1–34.

Deacon, T.W. (2010). What is Missing from Theories of Information? In P. Davies & N.H. Gregersen (Eds.), *Information and the Nature of Reality: From Physics to Metaphysics* (pp. 146–169). Cambridge University Press.

Campbell, D. (2024, June 8). London Hospitals Cancel Cancer Surgeries After Cyber-Attack. *The Guardian*. https://www.theguardian.com/society/article/2024/jun/07/london-hospitals-cancel-cancer-surgeries-after-cyber-attack. Accessed June 27, 2024.

Dunn Cavelty, M. (2008a). Cyber-Security and Threat Politics: US Efforts to Secure the Information Age. Routledge.

Dunn Cavelty, M. (2008b). Cyber-Terror—Looming Threat or Phantom Menace? The Framing of the US Cyber-Threat Debate. *Journal of Information Technology & Politics, 4*(1), 19–36.

Dunn Cavelty, M. (2013). From Cyber-Bombs to Political Fallout: Threat Representations with an Impact in the Cyber-Security Discourse. *International Studies Review, 15*(1), 105–122.

Dunn Cavelty, M. (2016). Cyber-Security and Private Actors. In R. Abrahamsen & A. Leander (Eds.), *Routledge Handbook of Private Security Studies* (pp. 89–99). Routledge, Taylor & Francis Group.

Dunn Cavelty, M. (2019). The Materiality of Cyberthreats: Securitization Logics in Popular Visual Culture. *Critical Studies on Security, 7*(2), 138–151.

Dunn Cavelty, M. (2020). Cybersecurity Between Hypersecuritization and Technological Routine. In E. Tikk & M. Kerttunen (Eds.), *Routledge Handbook of International Cybersecurity*. Routledge.

Durante, M. (2017). Ethics, Law and the Politics of Information: A Guide to the Philosophy of Luciano Floridi. Springer.

Dwyer, A.C. (2021). Cybersecurity's Grammars: A More-Than-Human Geopolitics of Computation. *Area.* https://doi.org/10.1111/area.12728

Dwyer, A.C., Stevens, C., Muller, L.P., Cavelty, M.D., Coles-Kemp, L., & Thornton, P. (2022). What Can a Critical Cybersecurity Do? *International Political Sociology, 16*(3), olac013.

Egloff, F.J. (2022). *Semi-State Actors in Cybersecurity*. Oxford University Press.

Egloff, F.J., & Dunn Cavelty, M. (2021). Attribution and Knowledge Creation Assemblages in Cybersecurity Politics. *Journal of Cybersecurity, 7*(1), tyab002.

Egloff, F.J., & Shires, J. (2021). The Better Angels of Our Digital Nature? Offensive Cyber Capabilities and State Violence. *European Journal of International Security,* 1–20.

Emerson, R.G. (2016). Limits to a Cyber-Threat. *Contemporary Politics, 22*(2), 178–196.

Eriksson, J. (2001). Cyberplagues, IT, and Security: Threat Politics in the Information Age. *Journal of Contingencies and Crisis Management, 9*(4), 200–210.

Farwell, J.P., & Rohozinski, R. (2011). Stuxnet and the Future of Cyber War. *Survival, 53*(1), 23–40.

Fisher, L., Palma, S., & Fildes, N. (2024, March 26). US and UK accuse China of Cyber Attacks on Politicians and Companies. *Financial Times.* https://www.ft.com/content/fc5d781b-47b9-461e-bbdf-b04229d895a3

Floridi, L. (2009). Philosophical Conceptions of Information. In G. Sommaruga (Ed.), *Formal Theories of Information: From Shannon to Semantic Information Theory and General Concepts of Information* (pp. 13–53). Springer.

Floridi, L. (2010). *Information: A Very Short Introduction.* OUP Oxford.

Floridi, L. (2013). *The Ethics of Information.* OUP Oxford.

Floridi, L. (2014). The Fourth Revolution: How the Infosphere is Reshaping Human Reality. OUP Oxford.

Floridi, L. (Ed.). (2016). The Routledge Handbook of Philosophy of Information. Routledge.

Fouad, N.S. (2022). The Non-Anthropocentric Informational Agents: Codes, Software, and the Logic of Emergence in Cybersecurity. *Review of International Studies,* 1–20.

Fuller, M. (2008). *Software Studies: A Lexicon.* The MIT Press.

Gartzke, E. (2013). The Myth of Cyberwar: Bringing War in Cyberspace Back Down to Earth. *International Security, 38*(2), 41–73.

Hansen, L., & Nissenbaum, H. (2009). Digital Disaster, Cyber Security, and the Copenhagen School. *International Studies Quarterly, 53*(4), 1155–1175.

Hare, F. (2009). Borders in Cyberspace: Can Sovereignty Adapt to the Challenges of Cyber Security? *Cryptology and Information Security Series, 3,* 88–105.

Hassib, B., & Alnemr, N. (2021). Securitizing Cyberspace in Egypt: The Dilemma of Cybersecurity and Democracy. *In Routledge Companion to Global Cyber-Security Strategy*. Routledge.

Healey, J. (2019). The implications of Persistent (and Permanent) Engagement in Cyberspace. *Journal of Cybersecurity*, 5(1), tyz008.

Henschke, A. (2021). Terrorism and the Internet of Things: Cyber-Terrorism as an Emergent Threat. In A. Henschke, A. Reed, S. Robbins, & S. Miller (Eds.), *Counter-Terrorism, Ethics and Technology: Emerging Challenges at the Frontiers of Counter-Terrorism* (pp. 71–87). Springer International Publishing. https://doi.org/10.1007/978-3-030-90221-6_5

Herzog, M., & Schmid, J. (2016). Who Pays for Zero-Days? Balancing Long-Term Stability in Cyber Space Against Short-Term National Security Benefits. In K. Friis & J. Ringsmose (Eds.), *Conflict in Cyber Space: Theoretical, Strategic and Legal Pespectives* (pp. 97–114). Routledge.

IBM. (2023). *Cost of a Data Breach 2023*. https://www.ibm.com/reports/data-breach

Jacobsen, J.T. (2020). Lacan in the Us Cyber Defence: Between Public Discourse and Transgressive Practice. *Review of International Studies*, 46(5), 613–631.

Jarvis, L., Macdonald, S., & Whiting, A. (2016). Analogy and Authority in Cyberterrorism Discourse: An Analysis of Global News Media Coverage. *Global Society*, 30(4), 605–623.

Junio, T.J. (2013). How Probable is Cyber War? Bringing IR Theory Back In to the Cyber Conflict Debate. *Journal of Strategic Studies*, 36(1), 125–133.

Kallender, P., & Hughes, C.W. (2017). Japan's Emerging Trajectory as a 'Cyber Power': From Securitization to Militarization of Cyberspace. *Journal of Strategic Studies*, 40(1–2), 118–145.

Kaplan, F. (2017). *Dark Territory: The Secret History of Cyber War*. Simon and Schuster.

Kaspersky. (2018) *Top 5 Most Notorious Cyberattacks*. URL: https://www.kaspersky.com/blog/five-most-notorious-cyberattacks/24506/. Accessed on June 15, 2020.

Kulikova, A. (2021). Cyber Norms: Technical Extensions and Technological Challenges. *Journal of Cyber Policy*, 6(3), 340–359.

Lacy, M., & Prince, D. (2018). Securitization and the Global Politics of Cybersecurity. *Global Discourse*, 8(1), 100–115.

Lawson, S. (2013). Beyond Cyber-Doom: Assessing the Limits of Hypothetical Scenarios in the Framing of Cyber-Threats. *Journal of Information Technology & Politics*, 10(1), 86–103.

Lawson, S.T. (2019). Cybersecurity Discourse in the United States: Cyber-Doom Rhetoric and Beyond. Routledge.

Lee, R.M. (2016). Protecting Industrial Control Systems in Critical Infrastructure. In R. Harrison, T. Herr, & R.J. Danzig (Eds.), *Cyber Insecurity: Navigating the Perils of the Next Information Age* (pp. 31–46). Rowman & Littlefield Publishers.

Lee, R.M., & Rid, T. (2014). OMG Cyber!: Thirteen Reasons Why Hype Makes for Bad Policy. *The RUSI Journal*, 159(5), 4–12.

Li, D., & Du, Y. (2017). *Artificial Intelligence with Uncertainty*. CRC Press.

Liebetrau, T., & Christensen, K.K. (2021). The Ontological Politics of Cyber Security: Emerging Agencies, Actors, Sites, and Spaces. *European Journal of International Security*, 6(1), 25–43.

Lindsay, J.R. (2013). Stuxnet and the Limits of Cyber Warfare. *Security Studies*, 22(3), 365–404.

McDonald, M. (2015). Contesting Border Security: Emancipation and Asylum in the Australian Context. In T. Balzacq (Ed.), *Contesting Security: Strategies and Logics* (pp. 154–168). Routledge.

McGraw, G. (2013). Cyber War is Inevitable (Unless We Build Security In). *Journal of Strategic Studies*, 36(1), 109–119.

McSweeney, B. (1999). Security, Identity and Interests: A Sociology of International Relations. Cambridge University Press.

Nye, J.S. (2017). Deterrence and Dissuasion in Cyberspace. *International Security*, *41*(3), 44–71.

Orman, H. (2003). The Morris Worm: A Fifteen-Year Perspective. *IEEE Security Privacy*, *1*(5), 35–43.

Rid, T. (2012). Cyber War Will Not Take Place. *Journal of Strategic Studies*, *35*(1), 5–32.

Rid, T. (2013). *Cyber War Will Not Take Place*. Oxford University Press.

Rid, T. (2016). *Rise of the Machines: A Cybernetic History*. W W NORTON & CO INC.

Russell, A., & Wang, H. (2002). How to Fool an Unbounded Adversary with a Short Key. In L.R. Knudsen (Ed.), *Advances in Cryptology—EUROCRYPT 2002* (pp. 133–148). Springer. https://doi.org/10.1007/3-540-46035-7_9

Salter, M.B. (2008). Securitization and Desecuritization: A Dramaturgical Analysis of the Canadian Air Transport Security Authority. *Journal of International Relations and Development*, *11*(4), 321–349.

Smith, S. (2005). The Contested Concept of Security. In K. Booth (Ed.), *Critical Security Studies and World Politics: Vol. Critical security studies* (pp. 27–62). Lynne Rienner Publishers.

Stevens, C. (2020). Assembling Cybersecurity: The Politics and Materiality of Technical Malware Reports and the Case of Stuxnet. *Contemporary Security Policy*, *41*(1), 129–152.

Stevens, T. (2015). *Cyber Security and the Politics of Time* (1st edition). Cambridge University Press.

Stritzel, H. (2011). Security, the Translation. *Security Dialogue*, *42*(4–5), 343–355.

Stritzel, H., & Chang, S.C. (2015). Securitization and Counter-Securitization in Afghanistan. *Security Dialogue*, *46*(6), 548–567.

Tarasewich, P., & Warkentin, M. (1999). Information Everywhere. In C.V. Brown (Ed.), *IS Management Handbook* (7th ed.). Auerbach Publications.

The Guardian. (2024, June 24). *NHS Confirms Stolen Data Published Online is From Blood Test Provider*. https://www.theguardian.com/society/article/2024/jun/24/nhs-england-confirms-theft-of-patient-records-data-from-its-provider. Accessed June 27, 2024.

The White House. (2003). *The National Strategy to Secure Cyberspace*. United States Government. https://www.us-cert.gov/sites/default/files/publications/cyberspace_strategy.pdf

Tsagourias, N., & Farrell, M. (2020). Cyber Attribution: Technical and Legal Approaches and Challenges. *European Journal of International Law*, *31*(3), 941–967.

Wilkinson, C. (2011). The Limits of Spoken Words: From Meta-narratives to Experiences of Security. In T. Balzacq (Ed.), *Securitization Theory: How Security Problems Emerge and Dissolve* (pp. 94–115). Routledge.

World Economic Forum. (2024, February 22). *3 Trends Set to Drive Cyberattacks and Ransomware in 2024*. https://www.weforum.org/agenda/2024/02/3-trends-ransomware-2024/

2 The Co-Production of Cybersecurity
Conceptualisation and Historical Overview

Introduction

When security theories attempted to widen the concept of "security" to include sectors other than the military, they focussed primarily on identifying those sectors' referent objects (i.e., threat targets or objects requiring protection), agendas, and logics of threats and vulnerabilities, rather than those sectors' *subject matter* (Buzan et al., 1998). This is understandable, since defining what the "military" is in military security, the "economy" is in economic security, or the "environment" is in environmental security is a reasonably straightforward task, given the long-standing resonance of those terms. The same does not apply to cybersecurity, however. Due to its novelty, technicality, and multidisciplinarity, what exactly the "cyber" is in cybersecurity remains quite vague, making the whole concept of cybersecurity comparatively far elusive than other security concepts. Any attempt to theorise this field, therefore, requires an exploration of the different ways it can be conceptualised and an answer to a very straightforward, yet paradoxical question: *what is cybersecurity?*

In many academic studies, the story of cybersecurity is often told as a fairly new one that dates back to the 1990s, when the term was first used in US policy circles, after being coined in a science fiction novel in 1984. Consequently, many literature take the 1990s as a starting point to trace the processes of securitisation of "cyberspace" and "cybersecurity" through political discourses (see Bendrath, Eriksson, & Giacomello, 2007; Branch, 2021; Dunn Cavelty, 2008; Hansen & Nissenbaum, 2009; Lobato & Kenkel, 2015). Implicitly, this suggests that cybersecurity initially emerged as a *non-security* sector, which then subsequently became discursively *securitised*. Although it is true that "cyberspace" and "cybersecurity" were novel terms at that period, their ontological status cannot be reduced to mere discursive utterances, or the ways they appeared in public discourses by multiple actors. If cybersecurity ultimately signifies the security of digital information systems, i.e., computers and networks, with all their associated software, hardware, and data – technologies that possess long historical roots – then such an ahistorical approach to studying its evolution would be both insufficient and over-simplified.

DOI: 10.4324/9781003454113-2

Accordingly, in answering the question *"what is cybersecurity?,"* this chapter presents a brief historical account of the evolution of computing and internetworking technologies – which constitute what we commonly refer to as "cyberspace" – from the Second World War (WWII) until the end of the Cold War. It argues that security has always been an integral part of the co-production of those technologies, as a context in which they were developed, as a generative force behind their evolution, and as a policy concern in different phases of their usage. Importantly, the chapter also shows how the historical development of "cyber" technologies, and their attendant security policies, were not entirely planned by their human inventors. As argued by some studies in Science and Technology Studies (STS) and digital humanities, information-based systems are more flexible and malleable than other technologies, and their usage seldom depends solely on their deliberate design. Those information artefacts are co-produced across a long period of time in cumulative processes, developing into "complex and recalcitrant textures," that are not just linked to their users' agency, but also to their own (Kallinikos, 2010). The chapter extends this argument to the construction and co-production of those systems' *security*. It demonstrates how computing and internetworking technologies as information systems evolved in ways that were not envisioned by their human inventors, which in turn influenced how their security has been conceptualised in different stages. This is primarily a process of *co-production*, in which human intentionality has been just one among many other constitutive elements.

Co-production is an idiom used to contextualise the production of scientific knowledge away from the deterministic, mono-causal approaches of its natural or social development. It is used in STS literature to analyse several aspects in the development of science and technologies, including: the emergence of new objects and their stabilisation, intelligibility and mechanisms of transporting ideas, and cultural practices that legitimise such ideas and assign specific meanings to them (Jasanoff, 2004). Although "co-production" is not explicitly an inquiry about agency, it can still enable us to broaden the concept of security away from the dichotomy of the human vs. the non-human. It allows for an understanding of technological development as a process of *intra-action* between discourses and materialities as two non-antagonistic constitutive forces (Jacobsen & Monsees, 2019). This chapter thus contends that the history of cybersecurity and "cyberspace" should be analysed as a complex process of restructuring, not just technically, but also politically and socially, in which the interests of various actors competed, and security considerations were intertwined with technical ones, and in many respects co-produced them.

To make this argument, the chapter starts with a brief historical exploration of the security context of the Cold War and how it influenced the development of science in general, and computing and internetworking research in particular. It highlights the generative influences of security considerations on the evolution of computers and the internet through the funding power of the military and its role in creating a market demand that shaped the supply of both technologies. The second section investigates the historical roots of computer and network security and how their conceptualisation had been evolving from concerns over physical security and unauthorised access to fears of malicious hacking and malware. Finally, the third section examines the current state of conceptualising cybersecurity, both on the academic and policy levels. It engages with other concepts that have strong links to cybersecurity, particularly ones that use information-based terminology, for conceptual clarification and delimitation. The chapter ends with an analysis of the cyber threat scope and the most significant policy challenges that dominate the current debate on cybersecurity.

Security as a context and a generative force for the co-production of "cyberspace"

In one form or another, security has always been an integral part of the development of "cyberspace": as a context in which it was developed, as a generative force behind many of its technologies, and as a policy concern in different phases of its evolution. Considering cybersecurity to be a completely new security challenge is problematic, since it overlooks the long history of interventions to achieve the security of computer networks and all the technologies associated with them (Ellis & Mohan, 2019). To illustrate this point, this section focusses on the evolution of computing and internetworking technologies since the end of Second World War until the advent of the internet. Although some roots of "cyberspace" or cyber technologies can still be found in the development of other electronic devices before the war, like punched cards (Heide, 2009), computers and networks are chosen as a starting point given their clearer links to the sort of modern cyberspace that we experience nowadays (Kello, 2017).

Studying the history and development of cyber technologies (computers and networks), along with their complex processes of co-production, cannot be done in isolation from the wider security context of the Cold War. That is because the Cold War had far-reaching implications on the course of scientific research as a whole, and the military-civilian partnerships that were formed to advance it. During that period, the Second World War was framed as a "scientific war" won by technological advancements achieved through the military's collaboration with academia, especially given the decisive role of the atomic bomb in ending the war and of the radar in winning it (Campbell-Kelly et al., 2014). There was a strong belief in both the USA and the Soviet Union that science can still win the Cold War for one of them. Consequently, advancement in science and technology became an integral part of both countries' national security strategies (Wolfe, 2013).

Together with the fear from an apocalyptic conflict with the Soviet Union (Chernus, 2008), this discourse contributed to raising the research and development (R&D) budget in the USA, even higher than war time, with the biggest share coming from the armed forces. Even after the National Science Foundation (NSF) was established in 1951 as a civilian institution to aid research, only 20% of computer research for instance was funded by it, while 50–70% received funds from the Department of Defence (DoD) (Edwards, 1997). That is what Eisenhower famously referred to as the "military-industrial complex" (Eisenhower, 1960, pp. 1035–1040), and others called the "military-industrial-academic complex" (Leslie, 1993).

One important institution that performed a significant role in this regard was the Office of Naval Research (ONR). Established in 1946, ONR was the first military agency to finance basic, unclassified research in academic and industry laboratories. Since it was the only federal agency to finance research immediately after the war, the ONR used its contractual authority to shape science policies, by selecting the fields, institutions, and individuals to be funded. Security imperatives were a major consideration for the ONR's contracts, particularly after the 1950s, with the rising congressional pressure to prove the relevance of research to national security and defence purposes (Sapolsky, 1990). The Advanced Research Projects Agency (ARPA) was another institution that influenced the post-war scientific research, particularly in fields like networking. ARPA was established in 1958, as a research agency affiliated to the DoD, following the surprising launch of the Sputnik satellite by the Soviet Union. Sputnik fostered the fear from a growing scientific gap that could allow the Soviets to attack the USA with ballistic missiles.[1] Consequently, ARPA was established with the responsibility of keeping the USA more technologically advanced than its adversaries and preventing any surprising events like Sputnik (Hafner & Lyon, 1998).

It could be argued that ONR and ARPA were products of existing processes of scientific and technological co-production, mediated in part by discourses of a scientific gap between the USA and the USSR, and fears of a Soviet attack. They were also a co-production agent that actively shaped the post-war scientific research per se, combining a complex set of scientific and military interests. This collaboration between the military and academia was legitimised by the security context of the Cold War. Many scientists were advisers to the government and advocated the increasing military spending on "basic research." Some even adopted a "two-title policy" for their research: a scientific title and a military-relevant one. An example for that was the scientific research on computational machines, which was re-oriented to be a research on command and control systems to suit military interests (Sapolsky, 1990).

The evolution of computers: from calculating devices to networked information appliances

The academic literature on the history of computers and networks is full of scientifically deterministic approaches that present an idealistic image of their development, by focussing on the success stories of their individual inventors (for example, see: Hafner & Lyon, 1998; Lavington, 2012). Few literature acknowledge the influence of the DoD and security considerations (for example:

Edwards, 1997). A closer analysis of the evolution of computers and networks reveals that their development path was largely unanticipated by their initial inventors. Computers started as calculating devices in the 1940s, and kept changing until they became a "networked information appliance" by the 1990s (Ceruzzi, 2003). The same applies to the internet, since the first network that resembles today's internet, the ARPANET, was designed initially to facilitate resource-sharing among academics, not interpersonal communications (Abbate, 1999). Yet, this should not lead to a conclusion that such unplanned processes were necessarily *accidental*, and consequently undermine the analysis of their socio-political context. As Jasanoff argues, "The design of technology is likewise seldom accidental; it reflects the imaginative faculties, cultural preferences and economic or political resources of their makers and users" (Jasanoff, 2004, p. 16).

Generally speaking, the history of computers can be linked to the evolution of information machines since the industrial revolution. The revolution brought about a growing need for information processing through technology, and resulted in the invention of typewriters, accounting equipment, desk calculators, and punched cards. Yet, the direct roots of modern computers can be more precisely found in the Second World War and the military's need for breaking adversaries' ciphers, creating tables for artilleries, and conducting ballistic calculations (Campbell-Kelly et al., 2014). From the 1940s until the 1960s, the US military was the main driving force behind technological developments in computers, not just through funding as part of the government's grand strategy during the Cold War, but also by being the main customer for computer products. Even though most computer research was conducted by universities and commercial laboratories, it was funded by the military and guided by its needs.

The first instance of the generative force of security in the development of computers goes back to the Colossus: the first electronic computer that was developed in 1943 by scientists contracted by the British government to decrypt adversaries' ciphers. Although the foundational idea behind digital computers was published in 1937, only the defence purposes of the war stimulated its application (Randell, 1982). The emergence of electronic machines gave rise to several computer-related disciplines, including cybernetics and artificial intelligence (AI). It was also transferred to other military applications, such as communication, intelligence, and command and control (Edwards, 1997).[2] Together with the Bombe – a machine developed to break the German Cipher device called "Enigma" – the Colossus sparked several computer projects outside the UK. This happened particularly in the USA, which took the lead in advancing computer research following the end of the war (Randell, 1982).[3]

Another machine that was generated out of the security imperatives of the war was the Electronic Numerical Integrator and Computer (ENIAC). The need for automated ballistic calculations for the army encouraged the development of ENIAC, financed by the Ballistic Research Laboratory (BRL), an army affiliate (Burks, 2014).[4] If it was not for the security needs of the war, the

ENIAC might not have been developed, as it was rejected and deemed as overly radical by the scientific community at that time (Flamm, 1988). The BRL continued to finance the development of the ENIAC into the EDVAC (Electronic Discrete Variable Automatic Computer); another military machine created to aid in a variety of tasks, including the development of a hydrogen bomb. The EDVAC marked the birth of internal programming, unlike the ENIAC which was externally programmed (Watson, 2012). The ENIAC and the EDVAC presented one big step in the history of government's support of "big science," especially that the amount of money these projects required was beyond the capacity of the private sector (Edwards, 1997).

Furthermore, the security needs of the military encouraged a sceptical private sector that did not initially acknowledge the importance of electronic computers, to get involved. For instance, the International Business Machines Corporation (IBM), which later dominated the commercial production of computers, was reluctant to enter the market until the military needs during the Korean war pushed it to develop a computer called "IBM Defence Calculator" or "701," sold to the military (Flamm, 1988). The ENIAC and EDVAC also gave boost to project Whirlwind, which was launched by Massachusetts Institute of Technology (MIT) to create a flight simulator for pilots training for the air force. Following the end of the war, it was integrated in the new computer-controlled air defence system, called the Semi-Automatic Ground Environment (SAGE). The sophistication of SAGE led to the development of various technologies that shaped today's computers, such as real-time computing, modems, and video and graphical displays (Agar, 2012).

Programming languages were another significant milestone in computer development, owing a lot of their success to the military. Two main programming languages were developed since the start of the 1960s: Cobol and Fortran. Fortran was introduced by IBM and proved successful because of IBM's domination of the market (Ceruzzi, 2012). On the other side, though it was a business-oriented language, the DoD not only pushed for the development of Cobol, it also encouraged its standardisation by announcing that "it would not lease or purchase a computer without a COBOL compiler" (Vee, 2017, pp. 108–109).

Gradually, the usage of computers started to broaden, as the technology showed potential for much more than military needs. Many computer amateurs began to view the idea of having affordable mini-computers for personal use as an important step towards human liberty. This encouraged several entrepreneurial start-ups to enter the market of personal computers, starting with Intel's development of the microprocessors, often regarded as "the enabling technology for personal computers." It was followed by the production of Altair 8000, the first version of a personal computer, in 1975. This accelerated the process of software production by various companies, such as Apple and Microsoft. Then computers became cheaper and affordable for non-military and non-business users, accompanied by the development of the World Wide Web (WWW) (Campbell-Kelly et al., 2014).

The evolution of the internet: from resource-sharing to communication platform

There are two main narratives in the literature on the development of the internet: one that presents it as an entirely technical project that evolved through spontaneous, unplanned processes (for example, see: Hafner & Lyon, 1998); and one that claims the network was developed just to maintain the robustness of military communication in case of a nuclear attack (for example, see: Siegel, 2008). Approaching the internet evolution as a co-production process, however, leads to a more complex conclusion that lies in-between the two narratives. The development of the internet can be described as an "organised chaos," produced by the overlapping interests of ARPA, NSF, programmers, developers, and even users (Marson, 1997). Both military and academic interests contributed to the evolution of a civilian internet, while corporate networking was initially sponsored by the state (Murphy, 2002).

Although the first network that resembles today's internet, the ARPANET, was not developed as a military network; the technology upon which it was based was generated out of security considerations. ARPANET was enabled by packet-switching, originally developed in the 1960s to secure the survivability of military communications by distributing them among different nodes to survive on redundancy in case of a strike (Ryan, 2010).[5] However, the radical nature of the idea hindered its immediate application. Only in the 1970s when ARPANET was created did packet-switching appear as a sound foundation for networking. ARPA financed the ARPANET project as the first version of a distributed network to allow its contracted academic institutions to remotely share their expensive computers (Brendon, 2001). Nevertheless, though it was not a military network, ARPANET was used for seismology and defence-oriented climate research. Also, it was not entirely an academic network, given the participation of the Army and the Air Force in it (Abbate, 1999).

Security needs also encouraged the application of packet-switching to two communication technologies used by the military in the 1970s: radio and satellites. A network called PRNET was established by ARPA to secure the military's command and control through radio packets; and another one under the name SATNET was created for the transfer of the military's seismic data. The existence of three heterogeneous networks – ARPANET, PRNET, and SATNET – and the need to connect them stimulated the development of the internetworking protocols TCP/IP as one cornerstone of the modern internet's architecture (Ryan, 2010). Furthermore, the military had an important role in the expansion of the network by obliging academic institutions contracted by ARPA to join it and mandatorily implement TCP/IP, despite the resistance of some to the idea of resource-sharing and internetworking (Ruttan, 2006). Additionally, the military's interest in a tightened security system for authentication and information-sharing was a first step towards the civilian internet. This was done by breaking the network down into two separate ones: a military network, MILNET, where a strict security system was implemented, and ARPANET continuing as a research network, and thus facilitating its expansion in the 1980s (Abbate, 1999).

The final and most critical step in opening the internet to the public was the privatisation of its backbone and the permission of its commercial use. The internet privatisation was neither easy nor inevitable, and was influenced by several technical, social, and political aspects, as well as the security considerations of the Cold War. It was facilitated by the US national security strategy, and the perceived scientific gaps between the USA and other international actors. This period witnessed wide congressional debates on funding supercomputers and networking, particularly in 1982, after Japan launched a project to supersede the USA in AI research. Consequently, data networking became an integral part of the national policy agenda, and the Congress endorsed public access to the internet. All of these facilitated the process of privatising the internet's backbone, which was completed by 1993 (Abbate, 1999).

In fact, many of the current debates on internet governance and security have their roots in this privatisation process, since it was not very carefully planned. For instance, following the privatisation, the NSF ruled out the backbone commercial providers from the jurisdiction of the Telecommunication Act, which regulated the activities of traditional communication carriers. The NSF did not put in place an alternative framework to prohibit discriminatory behaviour by those carriers. The absence of such framework resulted in many problems that are evident in the current struggle over net neutrality in the USA. In addition, the NSF did not impose any security requirements on backbone commercial providers, nor did it put any alternative arrangements for the security of the civilian internet after separating it from the military network. Thus, it could be argued that many internet security challenges are not the result of security being an afterthought in the development of the internet, but of the absence of an adequate preparation for the transfer of the internet's administration to commercial providers and for scaling it up (Shah & Kesan, 2007).

To summarise, the internet was neither just an academic project developed purely out of the enthusiastic ideas of its creators, nor was it entirely a military-oriented invention. The military's support, both in providing funds and creating a market demand, proved crucial in the evolution of the internet, starting from the early days of ARPA's adoption of packet-switching, to the development of the new technologies of satellite and radio switching. This all happened in spite of a sceptical academic community. Moreover, the diversity of the military operations produced a philosophy of decentralisation and heterogeneity in dealing with the network, as opposed to the industry's sponsored centralisation (Hicks, 1998). Hence, the co-production of the internet combined "the desire for cutting-edge research, openness of information, and adaptability to military needs" (Abbate, 2001, p. 150).

Security as a policy concern: from computer and internet security to cybersecurity

As a policy concern, the security of computers and networks has long historical roots that can be traced back to the introduction of the very first computer, even before the emergence of networks, malicious hacking, or malware. At each stage, this security was conceptualised differently, with diverse threat perceptions, referent objects, and utterances that reflect the technology's development state. In its early stages, computer security was shaped by the military, the same way computer research and industry were influenced by its interests. The DoD played an influential role in setting the security standards that governed early computer systems. This role diminished later with the increasing involvement of corporate actors, and even individuals, in the field of computer security. In these initial stages, and before the evolution of internetworking, computer security was mainly centred on three main issues: physical security, unauthorised access, and software bugs. Such threats defined what constituted a "computer crime" at that time, upon which legal frameworks were shaped. This conceptualisation of security reflected a belief in the controllability of machines, i.e., threats are external to machines and are calculable and controllable, and therefore defendable.

Physical security, unauthorised access, and software bugs

Following the end of the Second World War, the military was concerned about the security of its computer systems against physical attacks, theft, or natural disasters. As a result, computers' physical security became an integral part of the general security of military installations. Among the early perceived threats were the electronic radiations emanating from mainframe computers that allowed spies to decipher communications over computer systems (Yost, 2007). To deal with the issue during the 1950s, the government announced its first security standards for emanation levels, called TEMPEST, and the DoD obliged vendors to abide by them before they can sell any computer equipment to the government (Russell & Gangemi, 1991). Computers were also surrounded by containers to act as physical shields. Computer sabotage and manipulation did exist at that time, but were mainly physical and performed by insiders (Brenner, 2007).

The growth of time-sharing in the late 1960s, in which multiple users can share computer resources simultaneously, was perceived as an additional threat that could intensify unauthorised access. Therefore, RAND Corporation and the System Development Corporation (SDC) prepared what was known as the 'penetration studies' to identify vulnerabilities in time-sharing and resource-sharing systems. Additionally, a task force was established by the Defence Science Board in 1967, combining RAND researchers, defence contractors, scientists, NSA, and CIA officials, to examine ways through which the computer security standards of military environments could be applied to open environments. This task force produced a report in 1970 to examine the responsibilities of the computer industry in producing secure systems, which was also the birth of many cybersecurity terminologies that are used nowadays, such as vulnerabilities, threats, trap-doors, dependabilities, certifications, among others (Meijer et al., 2007).

With resource-sharing also came concerns over data privacy, which constituted what computer security was all about in technical literatures at that time (for example: Ellison, 1978; Wiesel, 1973; Reddy, 1979; Winkler & Danner, 1974). As a result, cryptography started to gain ground as an essential component of computer security, and the National Bureau of Standards announced, for the first time, the "national standards for cryptography."[6] Many companies, especially in the banking and petroleum sectors, invested heavily in encrypting their communications. For instance, IBM's spending on computer security reached 40 million dollars over a period of five years, and a major part of it was directed to cryptography research (Yost, 2007). Another important security concern in that period was software bugs: errors in coding. Prior to 1965, programming was done in closed environments and only programmers had access to written software. But as more people were getting involved in the process of software design, program correctness or fixing buggy software became an independent field of computer security research in 1974 (Meijer et al., 2007).

Nevertheless, concerns over unauthorised access and software bugs did not overshadow physical security as a perceived threat to computer security, even with the rise of phone phreaking in the 1970s, or hacking into phone systems for conducting free phone calls.[7] This is because the majority of security breaches to computer systems back then were primarily physical attacks. Examples include: the 1970s bombing of computer sites at the University of Wisconsin and New York University, the 1973s shooting of a computer in an American firm during protests in Australia, and the destruction of an IBM computer at Vandenberg Air Force Base in California in 1978 (Easttom, 2011).

On the legal and policy side, there were some attempts to face the rising threat of computer crimes, defined at that time as any physical destruction or unauthorised access to computer systems. This includes the Federal Computer Systems Protection Act, introduced in the Congress in 1977.[8] Though it was not adopted, it marked the first step towards recognising the security aspects of computer systems by the legislature (Easttom, 2011). Computer security was also acknowledged by the General Accounting Office, as one of the legislative agencies that provide services to the Congress. It issued multiple reports to the Congress recommending methods to combat any act that involved alteration of computer hardware or software, especially in federal computer systems, by employees or insiders (For example, see: The Comptroller General of the United States, 1976a). Additionally, other reports suggested limiting the number of federal employees who deal with computer systems to overcome the threat of unauthorised access (Marion & Hill, 2016); and analysing the physical security of computers against threats of fire, bombing attacks, among others (see: The Comptroller General of the United States, 1976b). The first actual law to be adopted in this regard was the Florida Computer Crimes Act in 1978, which criminalised all forms of unauthorised computer access, even if it did not involve any malicious intents, which was seen as a radical move at that time (Casey, 2011).

Yet, despite this increasing attention from the government and security practitioners, very limited knowledge about computer security was available to the general public at that time. The majority of news articles discussing computer systems mainly referred to their use in various settings, such as criminal courts (*The Washington Post*, 1977a), or in exchanging information among law enforcement agencies (Babcock, 1977), not their security. Those very few articles that focussed on the security aspects of such systems were concerned with what they referred to as "white collar crimes" in businesses, or the misuse of computers by employees to achieve financial gains (Kramer, 1977, 1978). Very limited articles discussed topics like unauthorised access or data privacy (*The Washington Post*, 1977b).

Malicious hacking and malware

During the 1960s and 1970s, hacking was approached positively, as part of the process of developing computer technologies. Because computers were not commonly used and were only accessed by researchers or military personnel, hacking was not widely known outside the computer science community. Even when hacking took place, it was done by students or computer programmers for exploration purposes. It was not associated with any malicious intents and was not subject to any sort of punishment (Marion & Hill, 2016). However, the emergence of networking, the commodification of information, and the "digital fences" implemented in computer systems with increased privatisation created a "cyber e-capital" that required protection. Therefore, the hacking culture started to be met with resistance, and a distinction was made between hackers (explorers) and crackers (robbers) (Dyer-Witheford, 2002).

Similarly, since the 1950s, the idea of self-copying and self-replicating programs was associated with the process of system design and development. Viruses and worms were part of the architecture of internetworking, as an essential tool for testing and experimenting the network. For this reason, even with increasing reports about computer viruses in the early 1980s, many security experts did not take them seriously and thought they were no more than an "urban myth." It took time for them to realise that viruses can go beyond being experimental programs towards malicious usages. Many incidents starting the 1980s played this role of shifting the emphasis to malicious hacking and malware as a security threat to networks and computers.[9] This transformed security conceptualisation in the field towards threats emanating from the machine, characterised by uncertainty and incalculability.

Examples include the first malicious virus to ever be reported in 1981/1982 that infected Apple II computer system; the "Brain virus" that targeted Microsoft's DOS system in 1986 (Skoudis & Zeltser, 2004); Ian Murphy's, or Captain Zap, hacking into the AT&T computer system and changing the clocks (Brenner, 2007); and hacking into the Los Alamos National Laboratory and the Sloan-Kettering Cancer Centre and stealing medical records by a group called 414 (Watson, 2012). Arguably, the most important of all such incidents was the Morris worm in 1988, which is often referred to as the "first internet worm" (Orman, 2003). This self-replicating worm, designed by Robert Morris, had a major influence in bringing internet security to the forefront of public attention and opening the doors in front of denial of service attacks

(Meijer et al., 2007).[10] It exposed the security vulnerabilities in the network, disabled many connected systems, and infected an estimated number of 6000 computers. Following this incident, a Computer Emergency Response Team (CERT) was established by the DoD, with the responsibility of coordinating responses in case of attacks, preparing security reports, and raising users' awareness about computer and network security (DeNardis, 2007).

As those operations were becoming more frequent and sophisticated, legislations to counter them were also developing. The first federal legislation on cybercrimes under the title "Computer Fraud and Abuse Act" was issued in 1986. Before this law was enacted, courts used to apply traditional laws on computer crimes with flexible interpretations. It was also preceded by some other efforts, like the amendments of the Comprehensive Crime Control Act in 1984 to include crimes of unauthorised access to computer systems; efforts that did not sufficiently respond to the rising challenge of computer crimes. The 1986s legislation protected computers of federal entities, financial institutions, and foreign commerce entities, against any unauthorised access of national security information, financial records, or any information held in federal departments or consumer reporting agencies. Subsequently, the law was used for the first time in the conviction of Robert Morris in 1988 and Herbert Zinn in 1989, who broke into the DoD computer systems (Easttom, 2011).

Several publications in that period mirrored this rising concern over computer and internet security, such as *The Orange Book*: a report published by the DoD in 1983, creating a common language for communication over computer security (Yost, 2007). The academic literature of the 1980s also revealed this shift from physical security towards software and hardware vulnerabilities, and the belief in the uncontrollability of machines (Ames, 1980; Fine, 1982). Network security, in addition to computer security, started to be emphasised in multiple literature (Kak, 1983; Rutledge & Hoffman, 1986); as well as the security of particular networks, such as the military (Landwehr, 1981; Stillman & Defiore, 1980); and the challenges facing hyper systems (Konheim, 1981; Oberman, 1983). By late 1980s, more literature began to discuss the increasing risks of viruses and worms to network security (Gardner, 1989; Joseph, 1988).[11] Furthermore, there was massive news coverage on computer and network security in the 1980s. Computer security made headlines in various news articles, discussing system vulnerabilities (Bakke, 1983; Burnham, 1985; Doyle, 1984; Francisco, 1982); rising computer crimes (Mc Cue, 1983); and the insider threat (Fitzgerlad, 1984). They called for several measures to tighten computer security (Ahern, 1982; The Associated Press, 1984), including the need for skilled computer specialists (Palma, 1980), good management (Hancock, 1981), and increasing public awareness (Byles, 1988).[12]

In conclusion, it could be argued that the military had strong influence on advancing computing and internetworking technologies. But these technologies later evolved in complex, unplanned processes into their modern versions and usages, beyond the intentions of their creators. Further, security was not a political discourse imposed on "cyberspace." The security of cyber technologies is as old as those technologies themselves and can be considered an integral aspect of their evolution. This, in turn, moves the question from whether cybersecurity is securitised or not, towards an investigation of how and why it is securitised and in what contexts.

The current state of conceptualising cybersecurity

Although computing and networking technologies evolved historically within the realm of security, the same does not apply to the concepts of "cybersecurity" and "cyberspace." The "cyber" terminology did not emerge in a security context and there was no intention for it to be used in security policy-making processes. This produced a condition in which these two terms are used differently in different contexts, with little to no agreement on what they really imply or include (Futter, 2018). The wide disagreement on conceptualising cybersecurity and defining its core referent objects extends both to academic and policy circles.

Conceptual delimitations

Cybersecurity, as argued by Dunn Cavelty, is a form of security that "unfolds in and through cyberspace" (Dunn Cavelty, 2013, p. 107). Before migrating to academic and policy debates, the term "cyberspace" first appeared in a science fiction novel called *Neuromancer* by William Gibson in 1984. The novel portrayed cyberspace as a "consensual hallucination" and a virtual environment that is not "real" (Betz & Stevens, 2013). The prefix "cyber" itself though is linked to cybernetics, which Norbert Wiener introduced in the 1940s as the science of control and communication in the animal and the machine, with a specific focus on human-machine interactions (Wiener, 1948). Following Gibson's novel, a "cyberpunk" culture started to develop, referring to a dystopian and futuristic style of writing. By mid-1990s, the term "cyber" began to be used in policy circles in the USA as a catch-all-phrase that encompasses what was previously classified as "information operations" and "information warfare." These operations included espionage, sabotage, communication fraud, and others that were revolutionised by informational technologies. But in Western discourses, the "cyber" terminology signifies a specific link to Computer Network Operations (CNOs): operations that have computer networks as both their attack tools and targets (Futter, 2018).

In academic literature, earlier demonstration of the concept of cyberspace reflected an understanding of it as synonym to the internet, and thus limited to its virtual manifestation (Bieber, 2000; Jordan, 1999; Loader, 1997; Mody, 2001). However, as the concept began to gain ground in the literature, some studies diverted from this narrow perspective towards defining cyberspace as a "construct" composed of multiple "layers." In this sense, cyberspace embodies a physical infrastructure (computers, cables, routers, and all hardware); a virtual layer, which is sometimes called "code" or "syntactic" level (programs, codes, protocols, and all software); and a "semantic" or cognitive layer (ideas and information stored in the system). Some literatures also add a "regulatory" level, of all the rules governing cyberspace, and a human level, related to individuals who deal with information systems (Applegate, 2015; Deibert et al., 2012; Libicki, 2007; McGuffin & Mitchell, 2014; Singer & Friedman, 2014). Accordingly, cybersecurity becomes all policies and tools undertaken to protect the cyber environment, with its virtual, physical, semantic, regulatory, and human components. Nevertheless, there remains broad disagreements on the relative importance of each of those "layers" and the nature of threats they should be defended against. As argued by T. Stevens, cybersecurity can be classed as a "wicked problem," as it is difficult to agree on which challenge to address and whether the solution implemented is not going to make the problem worse (Stevens, 2023).

Further, "cyberspace" is sometimes portrayed as just another space in which the dynamics of other security sectors can be detected. This is reflected, for instance, in the way cyber threats are commonly approached by adding the "cyber" prefix to conventional terms whose perception as threatening has long-standing resonance; such as cyber war, cyber espionage, cyber terrorism, and cybercrime (Carr, 2012; Goodman et al., 2007; Singer & Friedman, 2014). All such terms are widely used, albeit with little demarcation or clear definitions. In these formulations, any cyber-induced operation would be framed as an "attack" that falls under one of these categories. The problem with this attack-based conceptualisation is that it neglects mundane, everyday cyber threats and other cyber operations that are considerably influential on the long-term functionality of systems, such as cyber exploitation (Dunn Cavelty, 2016) and intelligence contests (Chesney & Smeets, 2023). As a result, a rather more pragmatic, technical approach is adopted by some scholars to classify cyber threats. Instead of using "cyber-attacks" as an all-encompassing category, they talk instead of CNOs that can take the form of computer network attack (CNA), computer network exploitation (CNE), or computer network defence (CND) (Brantly, 2014; Mazanec & Thayer, 2015). Rid also proposed classifying cyber threats into three categories: espionage, sabotage, and subversion (Rid, 2012).

However, some literatures are trapped in drawing comparisons between cybersecurity and other conventional sectors, in which the realities of those sectors control the perception of the cyber reality and determine its danger discourse. Some scholars assume that war, terrorism, espionage, and crime should have the same characteristics in cybersecurity as they do in conventional sectors for them to be conceptualised as such. For instance, they adopt Clausewitz's definition of war as violent, instrumental, and political to discard the possibility that war can take place in cyberspace (R.M. Lee & Rid, 2014; Limnell & Rid, 2014). According to them, only attacks that cause physical damage or massive destruction to the state's system can qualify as cyber terrorism (Stohl, 2007; Weimann, 2005). They assume conventional definitions of violence which are tied to lethality, and disregard how the serious implications of cyber threats on the social, economic, and political stability of societies may have transformed the concept of violence itself (Burton & Lain, 2020; Egloff & Shires, 2022). In assuming cybersecurity is a realm into which other strategic agendas are extended, the extent to which the logics and practices of cybersecurity may have actually transformed the concept of "security" as such goes often unexamined.

Hence, it could be argued that the prefix "cyber" has become a buzzword (Futter, 2018), and the term "cyberspace" has proven little relevance to socio-scientific analysis, even if widely used (Stevens, 2015). Given this ambiguity of the cyber terminology, cybersecurity can be easily confused with other related but different concepts. This may include internet security, which is only one aspect, or "layer," of what cybersecurity aims to protect. Another is internet governance, which is concerned with coordinating and managing the internet through IP addresses administration,[13] content regulation, copyrights protection, technical standards formations, etc. (Mueller, 2010). Although internet governance also encompasses security-related issues like protection against spam and cyber crime (Drake, 2005), this is just a subset of what cybersecurity is about. A more intricate relation exists between cybersecurity and all concepts that use information-based terminology. For instance, cybersecurity is sometimes used interchangeably with ICT security. The key difference is that ICT security is commonly used to signify the security of information infrastructure rather than information as such (here to mean data), while cybersecurity combines both (Von Solms & Van Niekerk, 2013). This can be counter-argued, however, by acknowledging that the security of information and that of the infrastructure it is stored on are predominantly inseparable.

It is also common for information security to be used as synonym for cybersecurity. Information security is defined in terms of the confidentiality, integrity, and availability of information (or data). Whereas cybersecurity is linked to computers and networks and takes place entirely in the digital realm, this is not necessarily the case in information security. The data that is the subject of protection in information security does not have to be stored on nor transmitted via digital devices; it can be written on paper and transmitted through verbal conversations, for example (Von Solms & Van Niekerk, 2013). Here, it is important to note that sometimes using the "cyber" or the "informational" language constitutes a political choice that reflects conflicting interests among actors. For instance, the Chinese and Russian governments use terms like information security, information weapons, and information warfare to replace the Western or Anglo-American language of the "cyber." According to them, information security is not reduced to operations that use or target computer networks; it also includes propaganda facilitated by mass media in order to influence the politics of a certain country (Giles & Ii, 2013). This distinction does not mean, however, that a certain operation cannot be both a CNO and an information warfare at the same time. An example for this is the hack against the US Democratic Party's national committee in 2016. This operation used and targeted computer networks, but also the data stolen was part of an information warfare to influence the elections (Futter, 2018).

Yet, disentangling the two concepts of information security and cybersecurity does not negate the informational ontology of the latter. Cybersecurity is not a synonym for information security, but it remains essentially informational. As defined by Dunn Cavelty, cybersecurity is

> a multifaceted set of technologies, processes and practices designed to protect networks, computers, programmes and data from attack, damage or unauthorized access, in accordance with the common information security goals: the protection of confidentiality, integrity and availability of information.
>
> (Dunn Cavelty, 2014, p. 89)

This definition makes it clear that cybersecurity is distinguished by the *digital* nature of the tools and targets of cyber incidents, but these tools and targets are primarily manifestations of *digital information*. This *informational* essence of cybersecurity will be explained and substantiated further in the next chapter. For now, it suffices to say that the limitations of the cyber terminology that the book responds to are not reduced to linguistic utterances; they are rather *theoretical* and *ontological*. Studying the informational ontology of cybersecurity is, therefore, not a call for replacing the cyber terminology with informational language in policy debates or academic research. Rather, attending to information is a bid to reach a level of theoretical abstraction for analysing the foundation of cybersecurity, beyond the discursive usage of particular terms.

Policy challenges

By reviewing academic literatures that deal with cybersecurity strategies and policy challenges, certain topics stand out as the most researched and widely debated, while others remain largely under-studied. As is the case with disagreements on conceptualisation, moreover, analysing cybersecurity strategies reflect varying understandings of the nature of "cyberspace" and cybersecurity and the relative importance of different types of cyber threats. In this regard, the cybersecurity literature explores some of the material conditions that affect cybersecurity, but mostly within a policy-oriented framework. Generally, this literature can be divided into two main themes: one that studies the strategic and operational challenges of cyber offence and defence; and one that focusses on cyber governance, state sovereignty, and public-private partnerships (PPPs).

Offence/Defence strategies and the cybersecurity dilemma

Many studies compare the relative usefulness of cyber defence and cyber offence as part of a cybersecurity dilemma that results in cyber insecurity for all states. Some argue that cybersecurity is offence-dominated. This is explained, for example, by the wide attack surface and the fact that easy accessibility to information technology lower the barriers of entry for would-be cyber attackers

(Clarke & Knake, 2010; Gjelten, 2013; Kello, 2013; Lynn, 2010). Nevertheless, other studies provide a counter-argument by analysing the advantages of cyber defence, and the possibility of using anonymity and deception effectively in defending against cyber threats (Gartzke & Lindsay, 2015). They point out the higher costs and shorter life cycle of offensive operations compared to defensive ones, in addition to the defence-favouring cybersecurity market (Rid, 2013). Further, other studies also assess the strategic value of cyber operations in peace and conflicts below the threshold of war, analysing their potential shortfalls and when they may succeed (Maschmeyer, 2021, 2023; Smeets, 2018), and investigating the barriers to entry into cyber conflict for states (Smeets, 2022).

One important aspect in this debate is the attribution problem (Egloff, 2020; Romanosky & Boudreaux, 2021), and its relation to cyber deterrence. The attribution problem refers to uncertainties in identifying the source of a cyber-attack, the kind of damages it caused to the targeted system, and the offensive capabilities of adversaries (Mazanec & Thayer, 2015; Egloff & Smeets, 2021). Together with the increasing developments in cryptography, these uncertainties make the application of traditional deterrence theory on cybersecurity quite problematic. However, some studies argue that deterrence is still possible and have already been done successfully in many cases (Rid & Buchanan, 2015; Klimburg, 2020). For instance, Nye argues that cyber deterrence can still be achieved through the threat of punishment; defence and resilience that deny the attackers advantage; interdependencies that raise the cost of the attack; and finally dissuasion by norms and threats to the state's soft power (Nye, 2017). This distinction between offensive and defensive operations is surpassed by other scholars who argue that cyber intrusions and exploitations can be used as a method of state defence. According to them, these "defensive operations," conducted mainly for espionage purposes, can be misinterpreted by the targeted states (Buchanan, 2016b), and thus render the offence-defence balance debate obsolete (Huntley, 2016).

Nonetheless, the implied assumption that the line between offence and defence is now blurred should be approached with caution. Although states may conduct cyber intrusions with defensive motives in the background, it is arguably problematic for academic literatures to deal with them as such. Just because an intrusion does not disrupt or cause damages to the target, does not mean that it is not offensive in nature. Such intrusions are still unauthorised and rely on the use of malware, which is commonly viewed as the "cyber weapon." Hence, some studies started to discuss the relationship between cyber espionage performed through CNIs on one side and international law and international cyber norms on the other (Georgieva, 2020; Hurel & Lobato, 2018; Sabbah, 2018; Mačák, 2017; Kello, 2021). This is important particularly given the complex interdependencies of cybersecurity and the fact that withholding information about the targeted vulnerabilities to be used by the state could eventually affect the security of everyone. This risk intensifies if we consider the materiality and contingent properties of information as such and the challenges they pose to the idea of human control implicit in such understanding of "cyber defence" – as will be further explained in the next chapters. For example, a malware used in targeting a certain system can spread to untargeted ones, even within the geographical location of the initiator, and result in several unintended consequences. Hence, there is a risk of condoning such operations by labelling them "defensive."

Cyber governance, state's sovereignty, and PPPs

Another approach to studying cybersecurity policy challenges is one that focusses on the issue of governance and the question of state sovereignty in a multi-stakeholder cyberspace (Cornish, 2015; Deibert & Crete-Nishihata, 2012; Shackelford, 2014). On one side, states try to maintain control over cybersecurity through militarisation, nationalisation, and other measures that may impede effective governance (Choucri & Clark, 2013; Fliegauf, 2016; Slack, 2016). For example, some studies criticise the growing tendency of many states to establish specialised military agencies for cybersecurity, also known as cyber commands, for their negative impact on democratic governance in fragile political settings (Solar, 2020) and their impact on digital human rights (Brantly, 2014; Halbert, 2016). However, on the other side, state sovereignty is also increasingly challenged in cybersecurity. One manifestation of this is cryptography, which relatively protects citizens' privacy, but at the same time provides anonymity to criminals and terrorists by encrypting their communications and facilitating network breaches (Moore & Rid, 2016). Encryption is even making corporations often "technically-prevented" from complying with states' demands for tracking users' activities in many cases (Buchanan, 2016a).

Relatedly, one of the most controversial issues in cybersecurity is the power dynamics between states and private actors, or the public-private nexus. Given the multi-stakeholder nature of cybersecurity, any successful strategy or policy would depend on effective partnerships between public and private actors. However, there are various impediments that complicate PPPs in cybersecurity, particularly in regards to managing CNIs. There is an incompatibility of interests between a state that perceives cybersecurity as a public good and a national security issue, and a private sector operating under profit-oriented business models. However, the power of private actors is growing to the extent that they can now influence processes of international norm creations in cybersecurity (Fairbank, 2019).

Challenges of PPPs crystallises in the information-sharing dilemma in cybersecurity. Early warning and attack alerts in cybersecurity normally depend on the first target's ability to first detect the attack, but most importantly, to share this information with the rest of potential targets. Information shared in these processes can include security policies, practices, and technologies, in addition to vulnerability information, liaison activities, anomalies, attack signatures, and attribution-related information. On one hand, governments are often reluctant to share vulnerability information with the private sector to protect its sources, and also to be able to exploit those vulnerabilities themselves in offensive or defensive operations. On the other hand, information-sharing is considered a business risk by the private sector. Sharing vulnerability or attack information risks financial losses, reputational harm, legal liabilities, or intellectual property rights theft, and may eventually affect profits (Carr, 2016; Dunn Cavelty, 2016; Dunn Cavelty & Suter, 2009; Etzioni, 2011; Muller, 2016).

Importantly, the public-private relations through which cybersecurity is constituted are more complex than the technicalities of sharing information. Specifically, they involve interactions that are not strictly "cooperative," and measures that can negatively affect human security. Edward Snowden's leaked NSA files, for example, revealed how the NSA exerted pressure on private companies for surveillance purposes, including Microsoft, Apple, Facebook, Google, among others, using court orders, withholding their licences, or hacking into their systems (Deibert, 2015). These measures altered the software and hardware of targeted devices, a process they called "interdiction" (Biham, Carmeli, & Shamir, 2016); weakened encryption by utilising supercomputers capable of cracking encryption algorithms; and enforced backdoor access to software (Harding, 2014). Added to this is the controversial involvement of the state in the black markets of vulnerabilities and exploits. Based on leaked reports, states are increasingly involved in such markets to build cyber arsenals to be used for offensive and/or defensive purposes. Such practices, in turn, undermine the long-term security of individual users; increase the market price of those vulnerabilities and exploits; and risk channelling information to the wrong hands (Dunn Cavelty, 2014, 2016; Herzog & Schmid, 2016).

The problem also extends to software manufacturers and the modes of production in the market. Many software vendors rush into introducing their products into the market with an intention to fix vulnerabilities later in the process, so that they can compete for profits. They mainly prioritise functionality over security, which led to a culture of acceptance of software insecurity. This commercialisation of cybersecurity transfers cyber risks and liabilities to the end user. Through the End Use License Agreements (EULAs),[14] vendors are transferring all risks and responsibilities to the end user, making themselves "bulletproof." The fact that many software is now free makes it difficult to hold those companies legally liable for any exploitable vulnerabilities in their products (Chong, 2016).

Against this background, the book will explore the intrinsic uncertainties and tendency towards disorder that characterise the operation of information systems as one challenge for cybersecurity. The challenges of information-sharing mentioned above will not be reduced to the practices of particular actors, be them state or private, but will also be linked to the properties of information as such. For example, the book will explore the ontological uncertainties of information systems that hinder the existence of particular information to share to begin with. This is extended not just to future unknowns, but even to present and past unknowns and *unknowables*. Entropic security, in this light, analyses cybersecurity challenges not *just* as a reflection of discourses or conflicting interests, but also as a product of the materialities of information per se.

Conclusion

This chapter has advanced a conceptually driven overview of cybersecurity, which engages with its definition, historical evolution, and major policy challenges. The chapter presented three main arguments. First, it countered a frequent tendency in the existing literature to view cybersecurity ahistorically. It showed how the roots of cybersecurity can be found in the development of computing and internetworking technologies following the Second World War until the internet was created – long before the terms "cyberspace" or "cybersecurity" even came into being. Secondly, and by way of extension, it demonstrated that security was not imposed on a purportedly non-securitised "cyberspace" through a set of exceptional political discourses. Rather, security has always been an important contextual influence, generative force, and policy concern in different stages of the evolution of all the technologies that constitute what we experience as "cyberspace." Thirdly, the chapter showed that both the historical development of these technologies and the evolution of their security conceptualisation were not entirely planned by particular actor(s). Rather, they evolved through a process of co-production, in which the interests of multiple actors competed, and their malleability generated new modes for their application.

As a context and a generative force, security was an indispensable element of the discursive and institutional tools of the co-production of scientific knowledge during the Cold War, which the internetworking and computing research was part of. Perceptions of the "scientific gap" and discourses that weaponised science were institutionalised, as in the cases of the ONR and ARPA, and subsequently utilised by those institutions in legitimising the influence the military exerted on scientific research, even on what academics regarded as "pure science." Furthermore, the foundational ideas that modern computers and the internet are based on, such as digital computations and packet-switching, were primarily generated by the war and post-war security needs, in a context of scepticism by both the scientific community and the private sector. Although computers started as calculators and the internet started as a resource-sharing network, both evolved into their modern versions and usages beyond the intentions of their creators.

As a policy concern, the security of computers and networks has always been present, though with different conceptualisation and diverse threat perceptions that reflect the technology's development state. Originally, when computers operated in controlled environments, there was an understanding of machines as necessarily controllable, and of threats as always extrinsic to them. That is why, computer security at that time was confined to unauthorised access and physical security. Yet, following the advent of networking and the dissemination of computers, the machines themselves began to be perceived as possibly threatening through software and hardware vulnerabilities. Therefore, malicious hacking and the use of malwares to target such vulnerabilities started to be viewed as a security threat upon which the modern cyber threat perception is based.

Though the current state of defining cybersecurity is marked by wide disagreements, it can still be defined as all the policies and tools undertaken to protect the multiple layers of the cyber environment, be it the virtual, physical, or semantic. Thus, concept like internet security or ICT security becomes subset of what cybersecurity is about. Despite the intersection between cybersecurity and information security, the earlier can still be distinguished by the digital nature of the attacks' tools and targets. Nevertheless, cybersecurity remains ontologically informational, or more precisely, *digitally* informational. Added to the conceptual ambiguity, there is a vast range of policy challenges that intensifies cyber insecurity and complicates the theorisation and study of cybersecurity. They include the cybersecurity dilemma and offence-dominance, PPPs, information-sharing, commercialisation of security, and governments' involvement in black markets of vulnerabilities and exploits.

Notes

1 The establishment of ARPA was also partially influenced by the Korean war (1953–1956), which led to lifting the spending cap Truman put on military research, and starting a remobilisation process of science to achieve military goals (Forman, 1987).
2 Computer science as a discipline first appeared in the 1950s as part of mathematics and electrical engineering departments. All through the 1960s, there were some challenges in defining what it is as a discipline and demarking and delineating it (P.E. Ceruzzi, 2003).
3 For more details on the comparison between the state of computing research in the USA and the UK and reasons behind the US lead, see (Bowles, 1996).
4 The ENIAC was not the first attempt to develop electronic computers; it was preceded by other efforts, not just in the United States of America, but also in Germany and the UK. However, these had little impact on the development of modern computers or the work of the Moore school for that matter. Additionally, the ENIAC was more complex and developed than all other electronic systems back then (Campbell-Kelly et al., 2014).
5 Packet-switching works by breaking up a message into a series of "packets," each with origin and destination labels. Those packets travel through the nodes in the network, choosing among multiple alternative paths based on efficiency or availability, until they are finally re-assembled at the destination. Thus, the disruption of any node does not impact communication's survivability, since packets can be always re-routed to an alternative path (Aksoy & DeNardis, 2007).
6 The National Bureau of Standards is a non-regulatory agency in the Commerce Department in the USA. It was later renamed The National Institute of Standards and Technology (NIST).
7 The most famous case of phone phreaking at that time was John Draper's or Capitan Crunch case. For more information, see: (Schwabach, 2014).
8 This bill is sometimes called the Ribicoff Bill after the senator who introduced it.
9 In 1983, a movie called *War Games* caught the public's attention by displaying the seriousness of hacking as a security challenge. The movie showed a teenager hacking into the computers of the North American Aerospace Defence Command, and almost triggering a third world war (Kaplan, 2017).
10 Denial of Service (DOS) attacks are those that flood a certain target with superfluous requests, which affects the target system's availability, by slowing it down or making it unreachable (Lee, 2013).

11 It is important to note here, however, that the vast majority of literature on computer and network security at that time were part of the computer science or engineering literature, not social sciences.
12 In 1988, the *Time* magazine published an important article that widely caught the public attention on the dangerousness of computer viruses. This article was exceptional because it portrayed an image of total cyber insecurity by using biological metaphors that compared computer viruses with epidemics (Elmer-Dewitt, 1988).
13 An IP address resembles a home address. Every device on a network has a unique set of numbers that identifies it, called the IP address, and which allows it to interact with other devices on the network (Panek, 2019).
14 EULAs are agreements shown to the user when using a software, containing all the rights and restrictions related to the software operation.

References

Abbate, J. (1999). *Inventing the Internet*. MIT Press.
Abbate, J. (2001). Government, Business, and the Making of the Internet. *Business History Review*, 75(01), 147–176.
Agar, J. (2012). *Science in the 20th Century and Beyond*. Polity.
Ahern, T. (1982, April 27). Experts Urge Computer Security. *The Associated Press*. https://www.nexis.com/
Aksoy, P., & DeNardis, L. (2007). *Information Technology in Theory*. Cengage Learning.
Ames, Jr. S.R. (1980). Security Kernels: Are They the Answer to the Computer Security Problem. *Wescon Technical Papers, 23*. Scopus.
Applegate, S. (2015). Cyber Conflict: Disruption and Exploitation in the Digital Age. In F. Lemieux (Ed.), *Current and Emerging Trends in Cyber Operations: Policy, Strategy and Practice*. Springer.
Babcock, C.R. (1977, June 17). Justice Puts Computer Plan on 'Hold'; Plan Raised Fears of 'Police State' Jurisdictional Crime Data Exchange. *The Washington Post*. https://www.nexis.com/
Bakke, B.B. (1983, April 25). Computer Security a Major Headache. *United Press International*. https://www.nexis.com/
Bendrath, R., Eriksson, J., & Giacomello, G. (2007). From 'Cyberterrorism' to 'Cyberwar', Back and Forth: How the United States Securitized Cyberspace. In J. Eriksson & G. Giacomello (Eds.), *International Relations and Security in the Digital Age* (pp. 57–82).
Betz, D.J., & Stevens, T. (2013). Analogical Reasoning and Cyber Security. *Security Dialogue, 44*(2), 147–164.
Bieber, F. (2000). Cyberwar or Sideshow? The Internet and the Balkan Wars. *Current History, 99*(635), 124–128.
Biham, E., Carmeli, Y., & Shamir, A. (2016). Bug Attacks. *Journal of Cryptology, 29*(4), 775–805.
Bowles, M.D. (1996). US Technological Enthusiasm and British Technological Skepticism in the Age of the Analog Brain. *IEEE Annals of the History of Computing, 18*(4), 5–15.
Branch, J. (2021). What's in a Name? Metaphors and Cybersecurity. *International Organization, 75*(1), 39–70.
Brantly, A.F. (2014). Cyber Actions by State Actors: Motivation and Utility. *International Journal of Intelligence and Counterintelligence, 27*(3), 465–484.
Brendon, L.K. (2001). ARPANET: An Efficient Machine as Social Discipline. *Science as Culture, 10*(1), 73–95.
Brenner, S.W. (2007). History of Computer Crime. In K.M.M. de Leeuw & J. Bergstra (Eds.), *The History of Information Security* (pp. 705–721). Elsevier Science B.V.

Buchanan, B. (2016a). Cryptography and Sovereignty. *Survival, 58*(5), 95–122.

Buchanan, B. (2016b). *The Cybersecurity Dilemma: Hacking, Trust and Fear Between Nations*. C. Hurst & Co Publishers.

Burks, A.W. (2014). From ENIAC to the Stored-Program Computer: Two Revolutions in Computers. In N. Metropolis (Ed.), *History of Computing in the Twentieth Century* (pp. 311–344). Elsevier.

Burnham, D. (1985, June 28). Lack of Security in Computers Seen. *The New York Times*. https://www.nexis.com/

Burton, J., & Lain, C. (2020). Desecuritising Cybersecurity: Towards a Societal Approach. *Journal of Cyber Policy, 5*(3), 449–470.

Buzan, B., Wæver, O., & Wilde, J. de. (1998). *Security: A New Framework for Analysis*. Lynne Rienner Publishers.

Byles, T. (1988, December 9). Public Awareness Vital to Computer Security. *Journal of Commerce*. https://www.nexis.com/

Campbell-Kelly, M., Aspray, W., Ensmenger, N., Yost, J.R., & Aspray, W. (2014). *Computer: A History of the Information Machine* (3rd edition). Westview Press, A Member of the Perseus Books Group.

Carr, J. (2012). *Inside Cyber Warfare* (2nd ed). O'Reilly.

Carr, M. (2016). Public–Private Partnerships in National Cyber-Security Strategies. *International Affairs, 92*(1), 43–62.

Casey, E. (2011). *Digital Evidence and Computer Crime: Forensic Science, Computers and the Internet*. Academic Press.

Ceruzzi, P. (2012). *Computing: A Concise History*. MIT Press.

Ceruzzi, P.E. (2003). *A History of Modern Computing* (2nd ed). MIT Press.

Chernus, I. (2008). *Apocalypse Management: Eisenhower and the Discourse of National Insecurity*. Stanford University Press.

Chesney, R., & Smeets, M. (2023). *Deter, Disrupt, or Deceive: Assessing Cyber Conflict as an Intelligence Contest*. Georgetown University Press.

Chong, J. (2016). Bad Code: Exploring Liability in Software Development. In R. Harrison, T. Herr, & R.J. Danzig (Eds.), *Cyber Insecurity: Navigating the Perils of the Next Information Age* (pp. 69–86). Rowman & Littlefield Publishers.

Choucri, N., & Clark, D.D. (2013). Who Controls Cyberspace? *Bulletin of the Atomic Scientists, 69*(5), 21–31.

Clarke, R.A., & Knake, R. (2010). *Cyber War: The Next Threat to National Security and What to Do About It*. HarperCollins.

Cornish, P. (2015). Governing Cyberspace through Constructive Ambiguity. *Survival, 57*(3), 153–176.

Deibert, R. (2015). The Geopolitics of Cyberspace After Snowden. *Current History, 114*(768), 9–15.

Deibert, R.J., & Crete-Nishihata, M. (2012). Global Governance and the Spread of Cyberspace Controls. *Global Governance, 18*(3), 339–361.

DeNardis, L. (2007). A History of Internet Security. In K.M.M. de Leeuw & J. Bergstra (Eds.), *The History of Information Security* (pp. 681–704). Elsevier Science B.V.

Doyle, J.J. (1984, February 20). "Hackers" Biggest Security Threat to Computer Industry. *United Press International*. https://www.nexis.com/

Drake, W.J. (2005). *Reforming Internet Governance: Perspectives from the Working Group on Internet Governance (WGIG)*. United Nations Publications.

Dunn Cavelty, M. (2008). Cyber-Terror—Looming Threat or Phantom Menace? The Framing of the US Cyber-Threat Debate. *Journal of Information Technology & Politics, 4*(1), 19–36.

Dunn Cavelty, M. (2013). From Cyber-Bombs to Political Fallout: Threat Representations with an Impact in the Cyber-Security Discourse. *International Studies Review, 15*(1), 105–122.

Dunn Cavelty, M. (2014). Breaking the Cyber-Security Dilemma: Aligning Security Needs and Removing Vulnerabilities. *Science and Engineering Ethics, 20*(3), 701–715.

Dunn Cavelty, M. (2016). Cyber-Security and Private Actors. In R. Abrahamsen & A. Leander (Eds.), *Routledge Handbook of Private Security Studies* (pp. 89–99). Routledge, Taylor & Francis Group.

Dunn Cavelty, M., & Suter, M. (2009). Public–Private Partnerships Are No Silver Bullet: An Expanded Governance Model for Critical Infrastructure Protection. *International Journal of Critical Infrastructure Protection, 2*(4), 179–187.

Dyer-Witheford, N. (2002). E-Capital and the Many-Headed Hydra. In G. Elmer (Ed.), *Critical Perspectives on the Internet* (pp. 129–164). Rowman & Littlefield.

Easttom, C. (2011). *Computer Crime, Investigation, and the Law.* Cengage Learning.

Edwards, P.N. (1997). *The Closed World: Computers and the Politics of Discourse in Cold War America.* MIT Press.

Egloff, F.J. (2020). Public Attribution of Cyber Intrusions. *Journal of Cybersecurity, 6*(1), tyaa012. https://doi.org/10.1093/cybsec/tyaa012

Egloff, F.J., & Smeets, M. (2021). Publicly Attributing Cyber Attacks: A Framework. *Journal of Strategic Studies, 0*(0), 1–32. https://doi.org/10.1080/01402390.2021.1895117

Egloff, F.J., & Shires, J. (2022). Offensive Cyber Capabilities and State Violence: Three Logics of Integration. *Journal of Global Security Studies, 7*(1), ogab028.

Eisenhower, D.D. (1960). *Public Papers of the Presidents of the United States, Dwight D. Eisenhower: Containing the Public Messages, Speeches, and Statements of the President, January 20, 1953 to January 20, 1961.* U.S. Government Printing Office.

Ellis, R., & Mohan, V. (Eds.). (2019). *Rewired: Cybersecurity Governance.* John Wiley & Sons.

Ellison, R. (1978). Does Computer Security Meet Privacy Requirements. *Information Privacy, 1*(1), 33–37.

Elmer-Dewitt, P. (1988, September 26). Technology: Invasion of the Data Snatchers. *Time.* http://content.time.com/time/subscriber/article/0,33009,968508-2,00.html

Etzioni, A. (2011). Cybersecurity in the private sector. *Issues in Science and Technology, 28*(1), 58–62.

Fairbank, N.A. (2019). The State of Microsoft?: The Role of Corporations in International Norm Creation. *Journal of Cyber Policy, 4*(3), 380–403.

Fine, L.H. (1982). The Total Computer Security Concept and Security Policy. *EDPACS, 10*(5), 1–20.

Fitzgerlad, J. (1984, February 3). Insiders Threat to Computer Security. *The American Banker.* https://www.nexis.com/

Flamm, K. (1988). *Creating the Computer: Government, Industry, and High Technology.* Brookings Institution Press.

Fliegauf, M.T. (2016). In Cyber (Governance) We Trust. *Global Policy, 7*(1), 79–82.

Forman, P. (1987). Behind Quantum Electronics: National Security as Basis for Physical Research in the United States, 1940–1960. *Historical Studies in the Physical and Biological Sciences, 18*(1), 149–229.

Francisco, S. (1982, March 2). Students Figure Way To Foil Computer Security. *The Associated Press.* https://www.nexis.com/

Futter, A. (2018). 'Cyber' Semantics: Why We Should Retire the Latest Buzzword in Security Studies. *Journal of Cyber Policy, 3*(2), 201–216.

G. Wiesel. (1973). Computer crime. Part 2: Data security. *Data Report, 1*(4), 24–27.

Gardner, P.E. (1989). The Internet Worm: What Was Said and When. *Computers and Security, 8*(4), 305–316. Scopus.

Gartzke, E., & Lindsay, J.R. (2015). Weaving Tangled Webs: Offense, Defense, and Deception in Cyberspace. *Security Studies, 24*(2), 316–348.

Georgieva, I. (2020). The Unexpected Norm-Setters: Intelligence Agencies in Cyberspace. *Contemporary Security Policy, 41*(1), 33–54.

Giles, K., & Ii, W.H. (2013). *Divided by a Common Language: Cyber Definitions in Chinese, Russian and English* (K. Podins, J. Stinissen, & M. Maybaum, Eds.; p. 17). NATO CCD COE Publications.

Gjelten, T. (2013). FIRST STRIKE: US Cyber Warriors Seize the Offensive. *World Affairs, 175*(5), 33–43.

Goodman, S.E., Kirk, J.C., & Kirk, M.H. (2007). Cyberspace as a Medium for Terrorists. *Technological Forecasting and Social Change, 74*(2), 193–210.

Hafner, K., & Lyon, M. (1998). *Where Wizards Stay up Late: The Origins of the Internet* (Touchstoneedition). Simon & Schuster.

Hancock, J.L. (1981, February 10). Management Involvement is Essential to Computer Security. *The American Banker.* https://www.nexis.com/

Hansen, L., & Nissenbaum, H. (2009). Digital Disaster, Cyber Security, and the Copenhagen School. *International Studies Quarterly, 53*(4), 1155–1175.

Halbert, D. (2016). Intellectual Property Theft and National Security: Agendas and Assumptions. *The Information Society, 32*(4), 256–268.

Harding, L. (2014). *The Snowden Files: The Inside Story of the World's Most Wanted Man.* Guardian Faber Publishing.

Heide, L. (2009). *Punched-Card Systems and the Early Information Explosion, 1880–1945.* JHU Press.

Herzog, M., & Schmid, J. (2016). Who Pays for Zero-Days? Balancing Long-Term Stability in Cyber Space Against Short-Term National Security Benefits. In K. Friis & J. Ringsmose (Eds.), *Conflict in Cyber Space: Theoretical, Strategic and Legal Pespectives* (pp. 97–114). Routledge.

Hicks, C.R. (1998). Places in the'Net: Experiencing Cyberspace. *Cultural Dynamics, 10*(1), 49–70.

Huntley, W.L. (2016). *Strategic Implications of Offense and Defense in Cyberwar.* 5588–5595. https://doi.org/10.1109/HICSS.2016.691

Hurel, L.M., & Lobato, L.C. (2018). Unpacking Cyber Norms: Private Companies as Norm Entrepreneurs. *Journal of Cyber Policy, 3*(1), 61–76.

Jacobsen, K.L., & Monsees, L. (2019). Co-production: The Study of Productive Processes at the Level of Materiality and Discourse. In M. Hoijtink & M. Leese (Eds.), *Technology and Agency in International Relations* (pp. 24–41). Routledge.

Jasanoff, S. (Ed.). (2004). *States of Knowledge: The Co-production of Science and the Social Order.* Routledge.

Jordan, T. (1999). *Cyberpower: An Introduction to the Politics of Cyberspace.* Routledge.

Joseph, H. (1988). Computer Viruses Can Be Deadly. *EDPACS, 15*(12), 1–6.

Kak. (1983). Data Security in Computer Networks: Guest Editor's Introduction. *Computer, 16*(2), 8–10.

Kallinikos, J. (2010). *Governing Through Technology: Information Artefacts and Social Practice.* Springer.

Kaplan, F. (2017). *Dark Territory: The Secret History of Cyber War.* Simon and Schuster.

Kello, L. (2013). The Meaning of the Cyber Revolution: Perils to Theory and Statecraft. *International Security, 38*(2), 7–40.

Kello, L. (2017). *The Virtual Weapon and International Order* (1st edition). Yale University Press.

Kello, L. (2021). Cyber Legalism: Why It Fails and What to Do About It. *Journal of Cybersecurity, 7*(1), tyab014.

Klimburg, A. (2020). Mixed Signals: A Flawed Approach to Cyber Deterrence. *Survival, 62*(1), 107–130.

Konheim, A. (1981). Guest Editor's Prologue. *IEEE Transactions on Communications, 29*(6), 761–761.

Kramer, L. (1977, November 18). Thieves, Swindlers Plague U.S. Business; Crimes Cost U.S. Business $30 Billion; Computer Bandits, Too. *The Washington Post.* https://www.nexis.com/

Kramer, L. (1978, June 22). Action Urged To Curb Crime By Computer. *The Washington Post*. https://www.nexis.com/

Landwehr, C.E. (1981). Formal Models for Computer Security. *ACM Computing Surveys (CSUR)*, *13*(3), 247–278.

Lavington, S. (2012). *Alan Turing and His Contemporaries: Building the World's First Computers*. BCS, The Chartered Institute.

Lee, N. (2013). *Counterterrorism and Cybersecurity: Total Information Awareness*. Springer Science & Business Media.

Lee, R.M., & Rid, T. (2014). OMG Cyber!: Thirteen Reasons Why Hype Makes for Bad Policy. *The RUSI Journal*, *159*(5), 4–12.

Leslie, S. (1993). *The Cold War and American Science: The Military-Industrial-Academic Complex at MIT and Stanford* (New edition). Columbia University Press.

Libicki, M.C. (2007). *Conquest in Cyberspace: National Security and Information Warfare*. Cambridge University Press.

Limnell, J., & Rid, T. (2014). Is Cyberwar Real: Gauging the Threats. *Foreign Affairs*, *93*, 166.

Loader, B.D. (Ed.). (1997). *The Governance of Cyberspace: Politics, Technology and Global Restructuring*. Routledge.

Lobato, L.C., & Kenkel, K.M. (2015). Discourses of Cyberspace Securitization in Brazil and in the United States. *Revista Brasileira de Política Internacional*, *58*(2), 23–43.

Lynn, W. J. (2010). Defending a New Domain: The Pentagon's Cyberstrategy. *Foreign Affairs*, *89*(5), 97–108.

Mačák, K. (2017). From Cyber Norms to Cyber Rules: Re-engaging States as Lawmakers. *Leiden Journal of International Law*, *30*(4), 877–899.

Maschmeyer, L. (2021). The Subversive Trilemma: Why Cyber Operations Fall Short of Expectations. *International Security*, *46*(2), 51–90.

Maschmeyer, L. (2023). Subversion, Cyber Operations, and Reverse Structural Power in World Politics. *European Journal of International Relations*, *29*(1), 79–103.

Marion, N., & Hill, J.B. (2016). *Introduction to Cybercrime: Computer Crimes, Laws, and Policing in the 21st Century*. Praeger.

Marson, S.M. (1997). A Selective History of Internet Technology and Social Work. *Computers in Human Services*, *14*(2), 35–49. https://doi.org/10.1300/J407v14n02_03

Mazanec, B.M., & Thayer, B.A. (2015). *Deterring Cyber Warfare: Bolstering Strategic Stability in Cyberspace*. Palgrave Macmillan.

Mc Cue, L.J. (1983, August 30). Computer Crime Fault of Security. *The American Banker*. https://www.nexis.com/

McGuffin, C., & Mitchell, P. (2014). On Domains: Cyber and the Practice of Warfare. *International Journal: Canada's Journal of Global Policy Analysis*, *69*(3), 394–412.

Meijer, H., Hoepman, J.-H., Jacobs, B., & Poll, E. (2007). Computer Security Through Correctness and Transparency. In K.M.M. de Leeuw & J. Bergstra (Eds.), *The History of Information Security* (pp. 637–653). Elsevier Science B.V.

Mody, S.S. (2001). National Cyberspace Regulation: Unbundling the Concept of Jurisdiction Note. *Stanford Journal of International Law*, *37*, 365–390.

Moore, D., & Rid, T. (2016). Cryptopolitik and the Darknet. *Survival*, *58*(1), 7–38.

The Associated Press. (1984, April 11). More Computer Security Needed. https://www.nexis.com/

Mueller, M.L. (2010). *Networks and States: The Global Politics of Internet Governance*. MIT Press.

Muller, L.P.N. (2016). How to Govern Cyber Security? The Limits of the Multi-Stakeholder Aproach and the Need to Rethink Public-Private Cooperation. In K. Friis & J. Ringsmose (Eds.), *Conflict in Cyber Space: Theoretical, Strategic and Legal Pespectives* (pp. 115–126). Routledge.

Murphy, B.M. (2002). A Critical History of the Internet. In G. Elmer (Ed.), *Critical Perspectives on the Internet* (pp. 27–48). Rowman & Littlefield.

Nye, J.S. (2017). Deterrence and Dissuasion in Cyberspace. *International Security*, *41*(3), 44–71.

Oberman, M.R. (1983). *Communication Security in Remote Controlled Computer Systems*. 219–227. Scopus.

Orman, H. (2003). The Morris Worm: A Fifteen-Year Perspective. *IEEE Security Privacy*, *1*(5), 35–43.

Palma, F. (1980, October 14). Need for Computer Security Specialists on the Rise. *The American Banker*. https://www.nexis.com/

Panek, C. (2019). *Networking Fundamentals*. John Wiley & Sons.

Randell, B. (1982). Colossus: Godfather of the Computer. In B. Randell (Ed.), *The Origins of Digital Computers* (pp. 349–354). Berlin, Heidelberg: Springer.

Reddy, Y. (1979). Data-Security in Computer-Networks. *Electronics Information & Planning*, *6*(12), 937–950.

Rid, T. (2012). Cyber War Will Not Take Place. *Journal of Strategic Studies*, *35*(1), 5–32.

Rid, T. (2013). *Cyber War Will Not Take Place*. Oxford University Press.

Rid, T., & Buchanan, B. (2015). Attributing Cyber Attacks. *Journal of Strategic Studies*, *38*(1–2), 4–37.

Romanosky, S., & Boudreaux, B. (2021). Private-Sector Attribution of Cyber Incidents: Benefits and Risks to the U.S. Government. *International Journal of Intelligence and CounterIntelligence*, *34*(3), 463–493.

Russell, D., & Gangemi, G.T. (1991). *Computer Security Basics*. O'Reilly Media, Inc.

Rutledge, L.S., & Hoffman, L.J. (1986). A Survey of Issues in Computer Network Security. *Computers & Security*, *5*(4), 296–308.

Ruttan, V.W. (2006). *Is War Necessary for Economic Growth?* Oxford University Press.

Ryan, J. (2010). *A History of the Internet and the Digital Future*. Reaktion Books.

Sabbah, C. (2018). Pressing Pause: A New Approach for International Cybersecurity Norm Development. In T. Minárik, R. Jakschis, & L. Lindström (Eds.), *CyCon X: Maximising Effects* (p. 20). NATO CCD COE Publications.

Sapolsky, H.M. (1990). *Science and the Navy: The History of the Office of Naval Research*. Princeton University Press.

Schwabach, A. (2014). *Internet and the Law: Technology, Society, and Compromises*. ABC-CLIO.

Shackelford, S.J. (2014). *Managing Cyber Attacks in International Law, Business, and Relations*. Cambridge University Press.

Shah, R.C., & Kesan, J.P. (2007). The Privatization of the Internet's Backbone Network. *Journal of Broadcasting & Electronic Media*, *51*(1), 93–109.

Siegel, P. (2008). *Communication Law in America*. Paul Siegel.

Singer, P.W., & Friedman, A. (2014). *Cybersecurity and Cyberwar: What Everyone Needs to Know*. Oxford University Press.

Skoudis, E., & Zeltser, L. (2004). *Malware: Fighting Malicious Code*. Prentice Hall Professional.

Slack, C. (2016). Wired yet Disconnected: The Governance of International Cyber Relations. *Global Policy*, *7*(1), 69–78.

Smeets, M. (2018). The Strategic Promise of Offensive Cyber Operations. *Strategic Studies Quarterly*, *12*(3), 90–113.

Smeets, M. (2022). *No Shortcuts: Why States Struggle to Develop a Military Cyber-Force*. Oxford University Press.

Solar, C. (2020). Cybersecurity and Cyber Defence in the Emerging Democracies. *Journal of Cyber Policy*, *5*(3), 392–412.

Stevens, T. (2015). *Cyber Security and the Politics of Time* (1st edition). Cambridge University Press.

Stevens, T. (2023). *What Is Cybersecurity For?* Policy Press.

Stillman, R.B., & Defiore, C.R. (1980). Computer Security and Networking Protocols: Technical Issues in Military Data Communications Networks. *IEEE Transactions on Communications, 28*(9), 1472–1477.

Stohl, M. (2007). Cyber Terrorism: A Clear and Present Danger, the Sum of All Fears, Breaking Point or Patriot Games? *Crime, Law and Social Change, 46*(4–5), 223–238.

The Comptroller General of the United States. (1976a). *Computer-related crimes in Federal programs: report to the Congress.* U.S. General Accounting Office. http://hdl.handle.net/2027/pur1.32754062639160

The Comptroller General of the United States. (1976b). *Managers Need to Provide Better Protection for Federal Automatic Data Processing Facilities, Multiagency: report to the Congress.* U.S. General Accounting Office. http://hdl.handle.net/2027/uiug.30112027352290

The Washington Post. (1977a, April 11). *Crime Computers.* https://www.nexis.com/

The Washington Post. (1977b, November 14). *HEW Computer Security.* https://www.nexis.com/

Vee, A. (2017). *Coding Literacy: How Computer Programming is Changing Writing.* MIT Press.

Von Solms, R., & Van Niekerk, J. (2013). From Information Security to Cyber Security. *Computers & Security, 38*, 97–102.

Watson, I. (2012). *The Universal Machine: From the Dawn of Computing to Digital Consciousness.* Springer Science & Business Media.

Weimann, G. (2005). Cyberterrorism: The Sum of All Fears? *Studies in Conflict & Terrorism, 28*(2), 129–149.

Wiener, N. (1948). *Cybernetics: Or, Control and Communication in the Animal and the Machine.* Wiley & Sons.

Winkler, S., & Danner, L. (1974). Data Security in the Computer Communication Environment. *Computer, 7*(2), 23–31.

Wolfe, A.J. (2013). *Competing with the Soviets: Science, Technology, and the State in Cold War America.* Johns Hopkins University Press.

Yost, J.R. (2007). A History of Computer Security Standards. In K.M.M. de Leeuw & J. Bergstra (Eds.), *The History of Information Security: A Comprehensive Handbook* (pp. 595–621). Elsevier Science.

3 Theorising Cybersecurity as an Infosphere

The Philosophy of Information Meets Security Studies

Introduction

In addition to being a "wicked problem" (Stevens, 2023), is cybersecurity also a peculiar security sector? Some literature attempted to theorise the arguable distinctiveness of cybersecurity through constructivist approaches that analyse the discursive peculiarities of this sector (Jarvis et al., 2016; Lawson, 2013; Lawson, 2019), of which the cyber securitisation literature is a prominent example (Bendrath et al., 2007; Dunn Cavelty, 2008a; Eriksson, 2001; Hansen & Nissenbaum, 2009). This book, however, moves from discourse to materiality by focussing on the *informational* ontology of cybersecurity. As will be demonstrated in this chapter, attending to *information* as a subject matter, referent object, and agency in cybersecurity adds important insights to its theorisation in Security Studies and to the understanding of the material conditions that influence its socio-political construction and its peculiarity.

The book introduces a novel analytical framework to the study of cybersecurity as an infosphere – a framework that goes beyond human subjectivity by also acknowledging the materiality of the informational *non-human*. This alternative framework is based on three main theoretical assumptions that the chapter explains and substantiates. Firstly, *cybersecurity is ontologically informational*. Information lies at the heart of all the technologies, sciences, and practices that enable this field and its very existence. Secondly, *information is a peculiar entity*. It is not just *another* non-human "thing"; rather, information has an autonomous status that sets it apart from matter and/or energy. Finally, it follows that *cybersecurity should be theorised differently* from other security fields. This is because information and its peculiar properties co-produce distinct meanings and logic(s) for "security" in cybersecurity, which requires new theoretical frameworks to understand.

In establishing these three hypotheses about cybersecurity, the chapter sets out a dialogue between the emerging field of the philosophy of information and Security Studies – particularly Critical Security Studies (CSS) which incorporates new materialist understandings of security. The chapter aims to prove that these different bodies of literature can bear on one another in providing a deeper understanding of cybersecurity than existing security theories. On one

DOI: 10.4324/9781003454113-3

hand, philosophy of information and information theory are employed to analyse the *ontology* of information as a peculiar referent object of security. On the other hand, new materialism is utilised to theorise information as a generative force in cybersecurity and for shifting from discursive performativity and human subjectivity towards a study of materiality. Together, these literatures contribute to an informational framework and a revised understanding of cybersecurity as entropic security, governed by the logics of negentropy, emergence, and noise.

The chapter unfolds in three sections. The first section explores the informational ontology of cybersecurity. Mobilising the philosophy of information, the chapter analyses the concept of information and demonstrates its intrinsic relationship to the sciences and technologies constitutive of the "cyber." The second section then draws upon new materialism to problematise the concept of agency in studying security. It establishes a genealogical link between new materialism and post-humanism on one side, and information and cybernetics on the other. The chapter then takes this new materialist argument about the agential role of non-human things a step further, by arguing that information too possesses distinctive agential capacities. Based on an understanding of information as the essence of the "cyber" and its peculiar properties, the third section moves to an argument that cybersecurity should therefore also be theorised differently. It focusses on the methodological aspects of theorising information as generative force of cybersecurity and of using discourse analysis in examining empirical data. In so doing, it also contests the security logic(s) of existentiality, exceptionality, and emergency that are often used to study cybersecurity, whilst simultaneously opening the way for the alternative logics of negentropy, emergence, and noise that will be further developed in the following three chapters.

The informational ontology of cybersecurity

Cybersecurity debates are marked by an abundance of objects requiring security protection, i.e., referent objects that cut across different sectors. As argued by Hansen and Nissenbaum, cybersecurity is better analysed through the "*competing* articulations of *constellations* of referent objects, rather than separate referent objects" (Hansen & Nissenbaum, 2009, p. 1163). The cross-sectoral connections in identifying referent objects can be seen in how the cyber threat is often conveyed in relation to the security of other sectors, particularly that of the military, the political, and the economic sectors. Since the traditional public/private and individual/collective divide is blurred in cybersecurity, it is common to find strong links among them in all threat discourses. The following statement in the US national cybersecurity strategy for 2023 is one example: "Cybersecurity is essential to the basic functioning of our economy, the operation of our critical infrastructure, the strength of our democracy and democratic institutions, the privacy of our data and communications, and our national defense" (The White House, 2023).

However, there is also something unique about cybersecurity that makes it an independent area of security policies and politics and not subsumed in other security sectors. There is a reason why we still call a hostile cyber operation that steals military secrets as a cybersecurity threat and not simply a military security one, even though it intersects with the latter. We categorise cyber espionage that targets the intellectual property rights of industries as primarily a cybersecurity challenge more than an economic security one. This is because the ultimate referent object of cybersecurity is *information*. Even if necessarily connected to the operation of the economy, military, and the daily lives of individuals, it is information that is the threat tool and object. It is information systems that are the immediate target of hostile cyber operations, and they are the object that cybersecurity measures seek to defend. But what exactly is information, and what makes it central to the ontology of cybersecurity?

There is, in fact, no widely agreed and clear understanding of what information really *is* – despite the abundant use of the term "information age," and the rapidly growing inventions of machines and processes to analyse, share, and store information, such as computers and networks. There are mathematical, algorithmic, physicist, biological, semiotic, and several other different approaches to define it in different sciences. It is considered a foundational concept for all of them, even if not accurately defined. This philosophical impasse the term has passed through in different stages of its history has been a hurdle towards developing a unified or universal theory of information (Deacon, 2010).

The first use of "information" as a scientific term dates back to Norbert Wiener and his work on cybernetics – which was defined as the science of control and communication in the animal and the machine and regarded as the "big bang of the information age" (Wiener, 1948). The birth of information theory, however, is usually associated with Claude Shannon's theory of communication that was introduced in 1948 (Shannon, 1948). Shannon was an American mathematician and electrical engineer who is regarded today as "the father of information theory" – with some even arguing that he "invented the information age" through his theory of communication (Soni & Goodman, 2017). Though a lot of the general philosophical exploration of information is based on their ideas, neither Wiener nor Shannon actually presented a definition or a theorisation of information per se. They rather dealt with information as a measurable quantity that they aim to maximise by minimising the noise in the transmission channel or medium (Burgin, 2010).

Nevertheless, there is a newly emerging field of research that interrogates the concept of information and introduces important philosophical insights about its nature and dynamics: the philosophy of information. The philosophy of information is a multidisciplinary field with contributions from physicists, mathematicians, computer scientists, linguists, biologists, among others. It evolved with the massive development of ICTs, and particularly computing and internetworking technologies that brought information to the centre of philosophy, as one significant force in the functioning of the world. The philosophy of information is broader and more inclusive than other approaches like digital philosophy, cyber philosophy, or computational philosophy that were also associated with the development of these technologies. As argued by Floridi, a prominent contributor to this new field, information philosophy is more concerned with information than computation because "it analyses the latter as presupposing the former" (Floridi, 2013, p. 15).

The multidisciplinarity of the philosophy of information as a field of research has already produced a wide variety of approaches to defining information. As noted by Floridi, information is "a polymorphic phenomenon and a polysemantic concept" (Floridi, 2009, p. 3). It has been regarded as "interpretation, power, narrative, message or medium, conversation, construction, a commodity, and so on" (Floridi, 2016, pp. 2–3). It can refer to the *process* of informing or changing the recipient's knowledge (information-as-process). In this sense, the intangible nature of the process means that information can be hardly measured. This process of informing is different from *information processing*, in which tangible entities, such as data and documents, are processed. Information may also denote *knowledge*, or the reduction of uncertainty (information-as-knowledge). It is sometimes used as an attribution of objects as "informative," be they documents or data (information-as-thing) (Buckland, 1991).

Alternatively, for the purpose of this book, information can be divided into three categories that speak directly to the field of cybersecurity: syntactic information, in the form of signs, signals, or bits; semantic information, or the meanings conveyed through those bits and signals; and pragmatic information, when those meanings and ideas conveyed are new to someone (Deacon, 2010). The syntactic definition is the basis for Shannon's theory of information, or the "mathematical theory of communication." According to him, the meaning that the signals carry is insignificant to the engineering problem of communication; i.e., information is totally a mathematical concept (Lombardi, 2016). For that reason, Floridi argues that Shannon's mathematical theory should be described as the "mathematical theory of data communication," or the study of the "syntactic level of information" (Floridi, 2009). On the other side, many scientists emphasise the significance of the semantic aspect of information: its meaning, reliability, and relevance. This semantic conceptualisation distinguishes information from data; assuming that data becomes information when meaning is added (Ratzan, 2004).[1]

Those multiple categories of information are central to cybersecurity. The syntactic layer is considered the "centre of gravity in cybersecurity," and is a fundamental quality that distinguishes cyber threats from conventional ones. All cyber threats must go through the syntactic layer to qualify as such; i.e., to originate from code alternations or the use of malicious codes (Friis & Ringsmose, 2016). Hostile cyber operations can also combine the semantic and the syntactic layers, as in the case of disinformation campaigns in which compromised computers – often called bots – are used to spread false information. Here, the meaning of information and its content is of absolute significance. Cyber incidents that aim at espionage, for example, are classified based on the semantics of the information they target, be it military secrets, personal identity, economic or business-related information, etc. Additionally, pragmatic information is fundamental for cybersecurity policies, since knowledge about vulnerabilities (exploitable coding errors) and cyber incidents is a key challenge. In consequence, many technical accounts of cybersecurity speak of it as "information assurance" instead, defined as "the art and science of securing computer systems and networks" (Ormes & Herr, 2016, p. 3). It is also typical to find "information security," or "infosec," in many technical studies used as a synonym to cybersecurity to strictly signify the security of digital systems (For example: Bishop, 2003; Dlamini, Eloff, & Eloff, 2009; Knapp, Franklin Morris, Marshall, & Byrd, 2009).

Further, information lies at the core of the philosophy of computer science, computer engineering, and software engineering, i.e., the sciences and technologies constitutive of the "cyber." This can be seen in some definitions of computer science as "the body of knowledge of information-transforming processes" or as simply "the study of information." This information can be data processing or methodological rules governing data structures (Primiero, 2016). Computing is also sometimes defined as "the systematic study of the ontologies and epistemology of information structures" (Primiero, 2016, p. 104). For that reason, it is argued that computation *is* an information processing and transformation process that operates by algorithms, or a set of instructional information. Instructional information is a type of semantic information that specifies actions to be performed by the receiver of information (Fresco & Wolf, 2016). The main function of a computer is thus the "material execution and mechanical realisation of those information-transforming processes" (Primiero, 2016, p. 91).

Correspondingly, although computers are syntactical devices, their operation is not restricted to syntactic information, but also involves semantics. For instance, a computing system used in an airplane has to operate with various natural semantic information about the condition of the airplane, including altitude, fuel injection, etc. (Piccinini & Scarantino, 2016). Information is also intrinsic to the physical layer of computational systems. Computer programs have their "syntax and construction rules" that establish control over the physical layer of computational systems, define their operation, and dictate how they should perform by executing certain actions. Such information can be referred to as "operational information." It creates a semantic relation between the ontology of the physical layer and the operational language of software (Primiero, 2016).

Here, it is important to make a clarification about the position of the *digital* in the analysis. As the previous chapter argued, cybersecurity can be distinguished from conventional definitions of information security by the *digital* nature of the tools and targets of incidents or threats. Why, then, does the book use "information" instead of "digital information" to theorise the ontology of cybersecurity? Typically, computation is always associated with the digital; it depends on manipulating digits, or variables with discrete/finite values (Piccinini & Scarantino, 2016). Also, digital computational processes combine all the previously mentioned elements of defining information:

the quantitative definition of bits, the syntactic construction of operations, the meaning of instructions, the abstract format of algorithm and
the epistemically loaded designer' intention.

(Primiero, 2016, p. 104)

Yet, the book does not confine its theoretical analysis to the digital because
there are many more insights that the non-digital literature on information can
add to the study – especially given that much of the "digital information" literature ignores semantic aspects that are also key in cybersecurity.

In fact, there are some philosophical and theoretical debates about the
connections between the analogue and the digital that may render an analysis
confined to the digital overly restrictive. Some physicists, for instance, argue
that any computer is "partly digital and partly analogue" (Dyson, 2001), or that
all digital devices are in essence analogical devices (Pias, 2005). For example, an
LCD screen can be considered a hybrid system between digital and analogue,
since it displays discrete pixels but emits light that may be measured using an
analogue continuum (Berry & Dieter, 2015). Engaging in this debate is beyond
the scope of this book; the main point here is that there is a theoretical and an
analytical sense in using information rather than *digital* information as a
framework, while acknowledging the digital nature of cybersecurity. There is a
lot that the general philosophical explorations on multiple forms of information can contribute to our understanding of cybersecurity and its peculiarities.
Additionally, using information rather than digital information as an analytical
framework avoids reducing information to a particular device or technology.

To summarise, if cybersecurity is to be defined as the security of computers
and networks, which are primarily information systems, then it may well be
argued that cybersecurity is fundamentally *informational*. This is ultimately an
acknowledgement of the informational essence of the *cyber*, its technologies,
and its sciences beyond linguistic utterances. Such an approach develops a
necessary level of abstraction that speaks to the fundamental being of cybersecurity. It also overcomes the ambiguity of the cyber terminology which has
done little to deal with the complexity of this realm by resorting to information
to understand its peculiarities. Investigating the informational ontology of cybersecurity ultimately renders it a study of materiality. As we will see below, this
approach challenges conventional security theories' conceptualisation of agency
and their focus on the discursive construction of security, whilst also speaking to
CSS attempts to incorporate new materialism or post-humanism in studying the
agential capacities of non-human objects in co-constructing security.

Agency and "information" that matters:

The idea that a non-human entity like information can be a generative force in
security construction coincides with what is often called "the material turn,"
"non-human turn," "thing studies," "post-humanism" or "new materialism" in
social sciences. This so-called turn produced new philosophical paradigms

such as object-oriented ontology (Bogost, 2012; Bryant, 2011; Harman, 2018), vital materialism (Bennett, 2009, 2015), agential realism (Barad, 2003, 2007), and actor-network theory (ANT) (for example: Latour, 2005; Law, 2002; Law & Singleton, 2005) to challenge the binary division of the world into human subjects and non-human objects. It is a call for widening the scope of research and philosophical debates away from the centrality of whatever is relevant to humans, be it reason, cognition, language, etc. (Kaltofen, 2018). Many of these contributions share a critical view of the dominant status of the Anthropocene, or the preoccupation with human subjectivities, but differ in the extent to which they move beyond this dominance in articulating the relationship between the humans and the non-humans (McDonald & Mitchell, 2017). They theorise matter as an active, politically significant force that has meaning beyond social, political, or economic structures. From this lens, the concept of agency, meaning the capacity to act or more generally "influence," in IR and Security Studies can be problematised.

In IR and Security Studies, the new materialist and posthumanist approaches represent a criticism of the inadequacy of existing ontological and epistemological perspectives to capture the non-human agency, be it that of machines, animals, bacteria, the environment, etc. For many years in the discipline, agency has been tied to the human subject, and the capacity to act has been linked to cognition, intentionality, desires, and decision-making – qualities regarded as exclusive to humans (Braun et al., 2018). Similarly, despite the CSS attempt to widen security to include actors other than the state, actancy was still limited to humans and human collectivities. If threats are conceptualised as manifestations of suffering, and if the ability to express such suffering is a function of humans, then security is tangled to human subjectivity (McDonald & Mitchell, 2017; Mitchell, 2014a, 2014b). However, a strand of research focussing on materiality and non-human agency started to gain momentum in recent years. Examples include the study of the materiality of critical infrastructure (Aradau, 2010), borders security (Bourne et al., 2015), weapons (Bousquet et al., 2017), emotions (Solomon, 2015), among others. Nevertheless, this approach is still not present enough on the mainstream research agenda, and there remains little agreement on what the "post" in a posthuman approach should look like (Cudworth et al., 2018).

Information and the evolution of the "material turn"

Information and cybernetics had a major impact on the evolution of thinking about human and non-human agency. If approached genealogically, post-humanism can be linked to the Macy Conferences on Cybernetics that took place between 1946 and 1953.[2] In these conferences, the human subject was decentralised in relation to non-human objects, and in particular to *information* (Wolfe, 2010). Cybernetics brought forward an analysis of information as a free-flowing entity among biological and non-biological systems, which opened the way towards blurring the lines between humans and machines.

The theoretical contributions of Wiener and Shannon in cybernetics and information theory resulted in thinking about humans as information processing entities; in turn, making them comparable to intelligent machines. Both humans and machines were seen as autonomous and goal-directed entities; an idea that challenges the humanist subject. As one of its pioneers, Ross Ashby argued that cybernetics is not concerned with "what *is* this thing?" but rather asks "*what does it do?*" (Ashby, 1958, p. 1).

In the Macy conferences, Wiener and John von Neumann (another influential mathematician) presented information as the most significant entity in the relationship between humans and machines. By approaching information as more fundamental and significant than matter/energy, analysing the similarities between humans, animals, and machines as informational entities became possible. In addition, Wiener's ideas about self-organisation in machines meant that they are able to self-evolve beyond human intentionality. This produced an understanding of self-evolving computer programs as "alive" themselves. Arguably, if everything in life is informational, then computer code is a "form of life" per se (Hayles, 2008). Accordingly, as argued by Mahon, "it is impossible to understand posthumanism properly without understanding cybernetics" (Mahon, 2017, p. 31).

Moreover, with increased digitisation, and the advancements in AI and robotics, the level of control humans maintain over machines began to be largely challenged (Cristiano et al., 2023). Such developments form the basis on which some futuristic trans-humanist approaches in social sciences conceptualise the "posthuman" and problematise the "human" as a category. From their perspective, humans are undergoing a process of evolutionary transformation towards becoming posthuman, i.e., being replaced, outpaced, and outsmarted by the technological non-humans. They contend that technologies are growing autonomously beyond human comprehension or control (Schwarz, 2017). For example, several studies on cyborgs, brain-computer interfaces, and biomedical engineering argue that the human body has transformed as a result of ubiquitous technological developments, either through upgrade, enhancement, extension, or invasion (Kaltofen, 2018).

However, acknowledging the co-productive capacities of technology and machines beyond human subjectivity does not mean they have overtaken agency. Unlike the techno-reductionist views of trans-humanism, the concept of agency can be problematised without resorting to biological, evolutionary, and hypothetical scenarios that assume humans are transforming into something else or being replaced entirely by the non-human. Rather, technology can be approached as one factor among many in breaking the binary division between humans and non-humans. This leads to what is often called a "flat ontology," in which the traditional separation between the human and the technological non-human is blurred, and thus necessitating a study of materiality (Braidotti, 2013; Salter, 2015). As such, the agency of non-human things is not that of intentionality, consciousness, or free will, but rather an agency of *influence*.

Such an approach, however, did not well migrate to the study of cybersecurity or to the field of IR in general. Technology is often approached either deterministically as a casual force or instrumentalised in studying the intentionality of a particular actor using a constructivist lens. Nonetheless, as argued by Bousquet, technology is less and more: "Less because it's not an external material agency that unilaterally transforms a passive social body, and more because it actually permeates every aspect of the social" (Bousquet, 2013, p. 96). The same applies to the theorisation of cybersecurity in the literature, where the logic(s) of security and risk are tied to human actors' intentionality or discursive practices. A new materialist or non-human revisiting of cybersecurity would thus require an analysis of information as an entity that *matters*. That being said, if all matter in all security sectors have co-productive capacities, the book argues that cybersecurity remains peculiar given its informational ontology. That is, *all matter matters, but information matters differently* – as will be shown next.

The peculiarities of information

In emphasising the agency of matter, many strands of new materialism include all non-human things in their entirety without distinguishing between the agential capacities they possess. In addition, some of them adopt a *relational* ontology, according to which things are only "real" if they have an effect on other objects. Latour's ANT, for example, assumes that there is no force embedded in things beyond their relations with other objects, from which they acquire agency (Latour, 2005). Nevertheless, it could be argued that "*all things equally exist, yet they do not exist equally*" (Bogost, 2012, p. 11). Matter or "things" are not all of one type or one form of agency (Bryant, 2014). As noted by Harman, flat ontology should not be an end in itself. It is not enough to reject the position of humans as the centre of ontology. Rather, the analysis should extend to investigating the different features and powers possessed by different "things." As he said, "an object is *more than its pieces* and *less than its effects*" (Harman, 2018, p. 53). Things can have a reality and intrinsic properties that are not necessarily reduced to their effects.

In the same vein, the book argues that information is a peculiar entity. It is not the mass-energy that scientific conceptualisation of "matter" signifies and not an ordinary non-human thing that is only important in relation to other objects. The roots of information's peculiarity can be found in Wiener's cybernetics. One central idea behind cybernetics is Wiener's argument that information plays a fundamental role in every aspect of the universe. According to him, all physical entities are inherently *informational*; i.e., all objects are "informational objects." He even argued that all living things, including human beings, are "informational entities" in how they store and process physical information in the form of genes, DNAs, proteins, etc. (Wiener, 1948). As a result, some studies argue that the information revolution has changed humanity more than any other revolution in history, because it influenced and changed objects at the "deepest level of their being," which is information (Bynum, 2016). The physicist John Archibald Wheeler took this argument even further in his article "It from Bit" by arguing that even matter-energy and all particles and physical entities (all *its*) owe their existence to information (to *bits*) (Wheeler, 1992).

Importantly, cybernetics presented information as different from matter and energy. Although it was not clear what Wiener meant by "information is information, not matter or energy" it can be inferred that he regarded information as "autonomous;" something that has a distinct structure (Janich, 2018). Information is much more complex and varied in its operation than matter or energy. It can be an objective abstract setting if approached mathematically, or a subjective entity that depends on a recipient. Although mainstream physics assumes a centrist position between objectivity and subjectivity in analysing the ontology of information, some physicists would argue that information is the only real thing in life. That is, matter and energy are just a reflection of information they embody, and the observer is just a subsystem that follows informational rules (Harshman, 2016). This argument is summarised by the information theorist Tom Stonier in saying:

> *Information exists.* It does not need to be *perceived* to exist. It does not need to be *understood* to exist. It requires no intelligence to interpret it. It does not have to have *meaning* to exist. It exists.
> (Italics in the original text; Stonier, 2012, p. 21)

Besides, information is distinguished by its inherent multiplicity: it has diverse sources, contents, and bearing media. It can be static (e.g., images or sentences), or dynamic (e.g., videos). It can be stored, transferred, modified, delayed, terminated, etc. It is used for communicative and non-communicative purposes (Sloman, 2011). Based on the receivers, it can be visual information, auditory, cognitive, etc. It can be *about* anything, from weather information, to political, economic, or military information (Burgin, 2010). It can be structural when imbedded in a system and kinetic when processed, transferred, or transformed (Stonier, 1991). It is temporal and ephemeral; its value changes over time and can also lose its value entirely at a certain point of time. It is not fungible since it does not have "identical interchangeable parts" and does not lose its components when given away. It is both a *process* and a *commodity*, and can be quantified in its syntactic form, or evaluated in entirely qualitative terms in its semantic form (Ratzan, 2004). It is "expandable," and its expansion is limitless. It is "compressible" and can be summarised or concentrated. It is "substitutable" and can replace many physical materials. It is "transportable," since it travels faster than physical objects and can be transmitted at speed of light or even faster, as argued by quantum mechanics. It is sharable, and although it does not decrease in quantity when shared, it may decrease in value according to the new recipient and their knowledge (Burgin, 2010).

Another very important quality of information is its transformational capacity. In one of its definitions, information is presented as "the difference that makes a difference." This entails both the capacity to *transform* and *change*. Information is always in a constant state of conversion. It is always changing, interacting with multiple other informational and non-informational agents, and transforming both itself and its surrounding environment. There are multiple ways through which the transformational capacity of information can be observed. For instance, information may transform during the process of its interaction with other information. In other cases, information may remain static while its perception by agents transforms. Alternatively, an agent may deliberately change information in an external, objective manner. But when the modified information changes the perception of another agent, the process turns into an external, subjective (or intersubjective) one (Gershenson, 2012). This transformational property links information to energy and leads some scholars to argue that energy may even eventually *"turn out to be information"* (Burgin, 2010, p. 102). That is why, information theory is sometimes portrayed as a framework that dictates the various types of transformations that can take place given available resources. It is also why computation is occasionally defined as a process of information transformation (Timpson, 2016), or assuming that information is only processed through data transformation (Burgin & Dodig-Crnkovic, 2013).

As a result, any self-organising system that is capable of making choices among many possibilities with the aim of achieving particular causal effect in its environment is regarded as one that generates information. Transforming inputs/causes into outputs/effects performed by self-organising system, unlike formalised mechanical systems, involves information generation. As one study argues, "the act to discriminate, to distinguish, to differentiate, is information" (Hofkirchner, 2012, p. 9). When information is generated, novelty and variety are also generated as a result of this constant transformation (Hoffmeyer, 2008). This variety and transformational capacity of information are important manifestations of its complexity. According to Ashby, in his work on complex systems and cybernetics, variety as such is a measurement of complexity: the more complex a system is, the more variety it has. He argues that information in itself is a "reflected variety," which is basically the difference perceived between one object and another (Hofkirchner, 2013). Similarly, Gregory Bateson – another key figure in cybernetics – argued that information and variety are synonyms (Hoffmeyer, 2008). Ashby called this the "Law of Requisite Variety," which contends that a system with complex variety requires models of management with an equal degree of variety to handle it: "only variety can destroy variety" (Grösser & Zeier, 2012, pp. 94–95). Put differently, complex information will result in more variety, requiring complex agents to perceive and manage it (Gershenson, 2012).

The complex nature of information is sometimes used to explain the complexity of the world despite the simplicity of the physical laws that govern it. This view presents an alternative explanation to cosmology and the origin of the world by replacing energy and matter with information as a constitutive force for evolution. That is to say, information, not energy, is the primary actor in the universe's history. As argued by Seth Lloyd, a professor of mechanical engineering and physics, if energy makes physical systems do things, information tells them *what* to do (Lloyd, 2006). Taking this argument a step further, some theorists see information as more fundamental than matter (Gleick, 2011). Such an argument replaces the traditional view of classical materialism that regarded matter – the material particles or atoms and their chemical interactions – as a force that explains everything in the world. Information, in turn, represents a new development in this cycle of transformation from matter to mass to mass-energy. This means that physical reality is better explained by information than the "mathematical relations" and the "laws of nature" that govern matter. Accordingly, information, not matter or energy, is the primary entity of change in the world (Mcmullin, 2010).

If we accept the definition of information as "the capacity to organise a system," then DNA, crystals, and almost everything that involves patterns of organisation or organised structures necessarily carry information. Even irregularities or chaos in some systems are sometimes the result of informational algorithms. This is because, as argued by Stonier, algorithms can be designed to have unpredictable results that may seem chaotic. Information is also implicitly found in all laws of physics. For example, motion is often understood as an "information act." Force and momentum are energy, while motion is information that specify the trajectory and structure of moving particles. Namely, moving particles carry energy and their motion carries information. The term "direction" can in itself also be described as "an information term." Basically, any change in time or distance denotes a change in the "information status" of a certain "moving body" (Stonier, 2012).

Derived from this viewpoint is an argument that the whole world is one giant computer, or quantum computer. This assumes that the universe is composed of bits, since all atoms and particles register "bits of information," and thus when the universe computes, it is basically computing itself. If computers are defined as machines that process information, then *anything* can also compute, including the universe (Lloyd, 2006). This argument belongs to a strand of literature referred to as "digital ontology" or "digital metaphysics." It primarily assumes that the ultimate nature of reality is informational and computational, and that the physical world can be explained by an "information-theoretic origin" (Steinhart, 1998). Even before the invention of computers or the idea of a digital computer, it was scientifically understood since the 19th century that "all atoms and elementary particles register bits of information," and when they happen to collide "those bits are transformed and processed." Yet, starting from the 1990s, it became clearer that those particles also *compute* and can be programmed. Hence, assuming that the universe is a computer is more than just a metaphor. The "computational theory of the universe" mentioned here has the power to explain complexity by reading the history of the universe as a history of subsequent "information-processing revolutions" (Lloyd, 2010).

The belief in the ontological primacy of information has echoes in some empirical studies that analyse the transformation of security and military conflicts in the informational age. For instance, Arquilla and Ronfeldt – who wrote the influential paper *Cyberwar is Coming!* (Arquilla & Ronfeldt, 1993) – view information as "an essential part of all matter," and that it is as fundamental to the world as matter and energy. Consequently, according to them, information "should be treated as a basic, underlying and overarching dynamic of all theory and practice about warfare in the information-age" (Arquilla & Ronfeldt, 1996, p. 154). Similarly, Dunn Cavlety and Brunner argue that information is *the* major source of power both in its material form of computers and infrastructure, and also in the "immaterial realm" of codes. As they put it, "Information becomes a weapon, a myth, a metaphor, a force multiplier, an edge and a trope – and the single most significant military factor" (Brunner & Cavelty, 2009, p. 633). Also, in his discussion of "network-centric warfare," Dillon contend that "information is the prime mover in military as in every other aspect of human affairs, the basic constituent of all matter" (Dillon, 2002, pp. 72–73).

However, it is important to note here that this ontological primacy of information is not an entirely resolved matter. Although this idea is adopted by most information philosophers, it is still contested by others, particularly some physicists. The physicist Julian Barbour, for instance, opposed Wheeler's argument "it from bit" by contending that the argument should be "bit from it." According to Barbour, information and "bits" cannot come to life without being underpinned in "things"; bits are merely dots on screens that are only given meaning by the physical universe, or by matter. For him, information is an *abstraction* not *reality*: "Try eating a 1 that stands for an apple" (Barbour, 2015, p. 204). This is a philosophical debate that McHarris sees as a resemblance of "the problem of the chicken and the egg" that drives the discussion into "an infinite loop" (McHarris, 2015, p. 233). Both sides of the debate, however, admit degrees of uncertainty about their argument. For instance, the physicist Anton Zeilinger says "What I believe but *cannot prove* is that quantum physics requires us to abandon the distinction between information and reality [emphasis added]" (Zeilinger, 2005). In criticising Wheeler, Barbour also acknowledges that his analysis "*weakens* but not necessarily *destroys* the argument that nature is fundamentally digital [emphasis added]" (Barbour, 2015, p. 197).

Though this remains an unresolved debate, it still signifies that information must be regarded as a highly peculiar entity. This peculiarity renders any approach that deals with information simply as *another* non-human thing or domain deeply problematic. To reiterate, *all matter matters, but information matters differently.* If the ontological status of information in relation to matter in the formation of reality is a debatable issue, it is not as such in the field of cybersecurity. As explained in the first section of this chapter, information is the core of all the layers that constitute "cyberspace" and cybersecurity and therefore its ontological fundamentality in this field is arguably more obvious. Once this informational ontology of cybersecurity is properly acknowledged, questions must therefore be raised about how the peculiarity of information impacts upon the construction and conceptualisation of security. The next three chapters, therefore, will each focus on one of three ontological properties of information that have the most profound ramifications for cybersecurity: (1) the intrinsic indeterminacies of information operation; (2) the complexities, non-linearities, and contingencies of syntactic information (i.e., codes/software); and (3) the simultaneous physicality and non-physicality of information. Using the aforementioned literatures on the philosophy of information, information theory, and software studies, the book will investigate how these three properties co-produce peculiar logic(s) or meaning(s) of "security" in cybersecurity. This all, however, first requires delineating a new methodological framework to the study of cybersecurity that is different from conventional security theories.

An informational framework to the study of cybersecurity

The realisation that information is a generative force in co-producing cyber-security discourses and practices challenges the position of actancy in conventional security theories and their application on the field. It is an argument that cybersecurity as an infosphere is constructed through the information *and* human actors. In this book, studying the generative force of information is illustrated in the different ways it challenges human control and intentionality in cybersecurity constructions, and produces different security logics that cannot be studied within traditional theoretical frameworks. As argued by Audra Mitchel's theorisation of posthuman security, instead of viewing the human as necessarily in control of security contexts, it is important to acknowledge the uncertainties and unpredictability that non-human objects produce (Mitchell, 2014b).

Further, the book gives more weight in the analysis to information, albeit without suggesting that it supersedes or replaces human agency. For example, in her introduction to vital materialism, Bennett says:

> I will emphasize, even overemphasize, the agentic contributions of non-human forces (operating in nature, in the human body, and in human artifacts) in an attempt to counter the narcissistic reflex of human language and thought. We need to cultivate a bit of anthropomorphism – the idea that human agency has some echoes in nonhuman nature – to counter the narcissism of humans in charge of the world.
>
> (Bennett, 2009, p. xvi)

Similarly, the book theorises information as generative and productive of the meaning of cybersecurity through its peculiar properties, alongside humans and in interaction with them, even though it focusses more on information as such. This analysis also entails that humans are not the only referent objects of cybersecurity, but information as such is a significant one too. Consequently, what matters for studying security is not the mere identification of referent objects, but a deep understanding of their ontology. That is what the book does by focussing on information. Instead of reducing objects to how we represent them or their relation to human subjects, we should investigate their own being (Bryant, 2011).

Additionally, the book introduces an analysis of cybersecurity that follows information as such (the entity), rather than cyber incidents (the phenomenon). It theorises information as such, rather than cyber-attacks or incidents, as generative for the meaning and practices of cybersecurity. The book also does not focus on particular political interventions to address cyber threats. Rather, it studies the overall meaning and construction of cybersecurity beyond a single incident or a specific intervention. That is to say, information matters, not only when a cyber incident takes place, *but also when it does not*. Further, the co-constitutive influences of information in cybersecurity are examined, not just in association with high-profile, state-backed cyber operations, but also in the case of the *mundane*, banal ones, or even when they do not take place at all.

This revised understanding of cybersecurity is done by establishing a dialogue between the newly emerging fields of the philosophy of information and software studies on one side and CSS and cybersecurity studies on the other. This dialogue is essential for any attempt to theorise cybersecurity in CSS given the close links that the philosophy of information and software studies have with the sciences and technologies constitutive of the "cyber." As argued by Bellanova, Jacobsen, and Monsees, security studies need to "take the trouble" of transcending disciplinary boundaries in order to understand the role of technologies through new frameworks of analysis (Bellanova et al., 2020). Transcending disciplinary boundaries is particularly important for theorising the materialities of cybersecurity given that, as argued by Tim Stevens, new materialisms have not sufficiently incorporated "information" as an entity in their "conceptual schema" and therefore did not engage with the important debates in the philosophy of information on the ontology of information in relation to the "matter" that new materialism theorised for (Stevens, 2012).

Discourse, materiality, and the non-human

Speech acts and discourses, traditionally regarded as exclusive to humans, hold a central position in how cybersecurity has been theorised in the literature, as explained before. They are both a signification of agency and a force that makes security possible. Hence, acknowledging the role of information in co-constituting cybersecurity necessitates contesting the focus on discursive practices in studying security. This is one of the major influences that the "material turn" had on IR: challenging the field's pre-occupation with meaning-making practices, textual and inter-textual analysis, and the question of representation (Lundborg & Vaughan-Williams, 2015).

Representationalism refers to a world view that ontologically and epistemologically distinguishes between "things" and "words"; "the observer" and the "observed"; and the "knower" and the "known"; i.e., the representations and the represented. In its essence, representationalism is concerned with accurately representing reality, or "correct correspondence" (Barad, 2007). When "things" and their representations are approached as separate, they are denied any capacity of influencing one another. The belief in the power of words to represent things is evident in some strands of CSS which adopt a wide conceptualisation of discourse to include the materiality of practices, but still "privilege words over things" (Aradau et al., 2014). For instance, some of the second-order securitisation literature criticised the theory for excluding contextual influences on securitisation processes (Balzacq, 2011; Wilkinson, 2011), but still left the non-human objects outside the realm of security. The contextual approach to security that this strand introduced to counter the linguistic focus of the original theory remain anthropocentric, i.e., deeply connected to humans and their perceptions and less attentive to the materiality of the non-humans that co-produce those perceptions.

In technology, software, and media studies, the term "materiality" is used in abundance, though rarely defined. In some instances, physicality is viewed as a defining character of matter, and therefore some studies refrain from using the term "materiality" when they discuss properties of software or data, for example, and use words like "stuff" instead (Leonardi, 2010). On the other hand, some STS literature use materiality to imply the social conditions that surround the development of technology and scientific discoveries; what could be described as "the materiality of practice." Similarly, in media studies, the materiality of the context is discussed in terms of the political economy or geographical considerations of media development, in addition to questions of ownership, control, reach, etc. (Dourish, 2017). Some of these aspects are also viewed from a political philosophy or a Marxian perspective, in order to analyse the material economic conditions of ICTs development that support their speed and manipulability, such as alliances between corporates and governments, international governance regimes, among others (Dourish & Mazmanian, 2013).

Nevertheless, acknowledging the *vitality* and *vibrancy* of non-human objects in new materialist scholarly contributions led to an increasing focus on materiality as a force of co-production in studying media infrastructures and digital technologies (Parks & Starosielski, 2015). An example of this approach can be seen in studies that attend to the cultural role of information and "digital goods" and their symbolic weight in material cultures, as well as the study of how information and digital networks are shaping *space* (Dourish & Mazmanian, 2013). This is the strand of research in studying materiality that the book primarily speaks to by analysing the generative capacities of information in co-producing the field of cybersecurity.

Materiality and discourse, however, should not be viewed as two opposing frames of analysis. Materiality (the non-human things that make up our existence) and discourse (meaning-making practices) can be both regarded as co-constitutive forces of security. For example, as explained by Claudia Aradau, the scanned image of a liquid that is checked separately in an airport for security concerns is co-produced by a *screening device* that identifies this liquid as a "distinctive object," and also by *discourses* on terrorism and precautionary measures. That is, the "dangerousness" of the liquid in this example is co-constituted by both materiality *and* discourse (Aradau et al., 2014). Barad also argues that materiality and discursive practices should be theorised in productive terms. This does not mean simply considering the role of materiality *in addition* to discourses, but most importantly, their *intra-action* or entanglement. That is to say, discursive practices are not a property of human actors; they are rather "material (re)configurings of the world through which boundaries, properties, and meanings are differentially enacted" (Barad, 2007, p. 183). Similarly, materiality is also discursive, and neither matter nor discourses have an ontological or epistemological superiority over the other. It is possible to connect words, things, and the material in understanding the complexity of "their becoming," without having to consider the human as either a cause or an effect (Barad, 2007).

By reconceptualising discourse and breaking its traditional link with linguistic utterances, we can then view non-human entities, or information in this book, as 'produced and productive, generated and generative' (Barad, 2007, p. 137) and as a generative force in the co-construction of its own (in)security. As Aradau argues in her work on the securitisation of critical infrastructure, the materiality of non-human things is capable of enabling and constraining what is securitised. Securitisation, she explains, should be studied as a process of *materialisation* in which the generative capacities of matter and "things" and the role of materiality in co-constituting reality are considered (Aradau, 2010). This means too that the effects of non-human things should not be reduced to human linguistic framings. Rather, discourse and materiality should be approached as agents in a process of co-production (Jacobsen & Monsees, 2019).

Drawing on these contributions, the book reconceptualises the discursive in analysing the policy documents and congressional hearings, as part of the empirical case study. Discourse is not approached as language or speech act, but rather as the force that enables or conditions language. Bridging the gap between the material and the discursive aims at analysing information not merely as an object of protection or a facilitating condition in cybersecurity, but as a generative force of its own (in)security. It follows that the process of "securitising" cybersecurity as an infosphere is in essence a process of materialisation. In such process, information plays a key role in co-producing the logic(s) of security and shifting them from conventional understandings.

The logics of security-risk

One important implication of employing an informational approach to study cybersecurity is challenging the fixation of security logics that characterise many literatures in Security Studies. For example, the securitisation theory, which has been applied extensively to theorise cybersecurity, ties security to existential threats and exceptional measures. A security threat only qualifies as such when it is linked to the survival of the referent object, and securitisation succeeds when "above-politics" measures are proposed and legitimised. These logics were criticised for consolidating state's power and sovereignty in the realm of emergency (M.C. Williams, 2003); for reflecting a "Cold War mindset" that restricts the theory to the statist logics of militarisation (Booth, 2007); for neglecting the political implications of securitisation; and for its liberal-democratic conception of "normalcy" (Aradau, 2004). As a result, some scholars think of security as undesirable and problematic (Balzacq et al., 2015). For instance, Didier Bigo defined security as some sort of a political technology or a "dispositif," in which exceptionalism and exclusion are institutionalised, normalised, and banalised, through processes of insecuritisation (Bigo, 2000, 2002, 2006, 2008). Therefore, many CSS criticise securitisation theory by rejecting the concept of security entirely and replacing it with alternative paradigms, such as emancipation (Aradau, 2004; Aradau & Van Munster, 2016; Booth, 1991, 2007).

Another way such logics are scrutinised and replaced can be found in the risk literature. Risk was introduced in Security Studies with the assumption that it has transformed security, either by replacing or intensifying it. The argument that risk is the new security is centred on the sociologist Ulrich Beck's idea of *risk society* (Beck, 1992, 1999, 2002, 2006). Beck's thesis assumes that we are now living in a "second modernity," whereby risks can be conceptualised as "a systematic way of dealing with hazards and insecurities induced and introduced by modernisation itself" (Beck, 1992, p. 21). Several security studies utilise Beck's ideas to argue that risk has changed the international security agenda (Rasmussen, 2001, 2004); shifted the focus of strategic studies to risk-based instead of threat-based security strategies (M.J. Williams, 2008); and replaced the security dilemma with a "security paradox" that deals with the management of uncertainty rather than the management of insecurity (Kessler & Daase, 2008). Others analyse the negative implications of this arguable replacement of security with risk. They argue that risk has strengthened states' sovereignty in exceptional ways (Aalberts & Werner, 2011; Stockdale, 2013), and has been commercialised and commodified by private companies to enhance their profits (Krahmann, 2011). Risk-based, as opposed to threat-based, approaches are also used to study some empirical cases of cybersecurity practices (Backman, 2023; Kaminska, 2021).

However, this strand of research was criticised by many in the CSS, arguing against its realist ontology and introducing a framework based on Foucault's concept of *governmentality*. By conceptualising risk as a "social technology," some CSS scholars contend that Beck's thesis fails to capture the diverse ways and mechanisms put forward to govern risk and to render the uncertain future knowable and actionable (Aradau et al., 2008). They see risk as part of a neo-liberal rationality and a *dispositif*, composed of multiple material and discursive elements, for governing what is perceived as ungovernable (Aradau & Van Munster, 2007). Assuming this inevitable association between the neo-liberal governmentality and risk has led to a fixed perception of the security-risk nexus. That is, many argue that risk intensifies and expands security, or acts as a force multiplier, by privileging decisions based on dangerousness, normalising exceptionalism, and reinforcing authoritative presentations of insecurity (Aradau et al., 2008; Aradau & Van Munster, 2007; De Goede, 2004; Hagmann & Dunn Cavelty, 2012; Salter, 2008).

The bottom-line is that fixing security and risk logics has led to an understanding of each of them as a distinct paradigm. This, in turn, resulted in an under-theorisation of the security-risk nexus in the study of cybersecurity. If security is assumed to be always existential and exceptional, many practices and discourses that are enabled by risk would be seen as inferior to the study of security. For instance, in her explanation for why cyber securitisation has failed, Dunn Cavelty argued that "Despite the fact that there is national security rhetoric in abundance, the actual countermeasures in place rely on risk analysis and risk management" (Dunn Cavelty, 2008b, p. 30). Similarly, although Hansen and Nissenbaum recognised that some of the cybersecurity peculiar grammars they presented resonates with risk theory, they did not incorporate risk in their study: "since 'security' rather than 'risk' is the dominant policy as well as academic concept, it is not pursued in further detail in this paper" (Hansen & Nissenbaum, 2009, p. 1164).

Fixing the logics of security has been criticised by some studies for contradicting the idea of intersubjectivity in security constructions. They view it as a prioritisation of the security analyst over the practitioner and an assumption that the former has better understanding of the essence of security than the latter (Ciută, 2009; Floyd, 2016). Others argue that the meaning of security is always subject to negotiations and constant challenging depending on the context, which may result in its eventual transformation beyond fixed logics (Stritzel, 2011; Trombetta, 2021). Similar arguments can be also made concerning fixing the logic and meaning of risk. Assuming that risk is inherently an un-securitised or a de-securitised articulation, as argued by Corry (2012), may lead to overlooking the ways through which risk has been used to facilitate and enable securitisation in some cases rather than having a desecuritising effect (for example, see: Elbe, 2008).

However, this book takes a different line of criticism to the fixation of the logic(s) of security and risk and their nexus in the study of cybersecurity. It argues that security and risk in the infosphere are not *just* what human actors make of them, and therefore their logics should not be tied exclusively to those actors' intentionality, perception, and representation. Emergency measures should not be perceived as subject only to human actor's intentions. As will be substantiated in the next chapters, even with the presence of an intentional actor and a desire to go beyond the "ordinary" or "normal" in cybersecurity, the materialities of information would lead to contingencies and uncertainties that challenge such actions. Similarly, the conceptualisation of existentiality – not its normative disposition – should be further interrogated. There remains a gap in determining what it means for a threat to be existential and analysing the link between existentiality on one side and immediacy, urgency, and the physical on the other. In application, many studies assume that it suffices to present a threat as "so severe" or serious for it to be considered existential (for example: Collins, 2005). By extension, whether digital information can be existentially threatened or constructed as such is an important question that the cybersecurity literature did not sufficiently engage with. In short, adopting an approach that considers the materiality of information reveals the problems of imposing conventional security logics on cybersecurity.

Conclusion

This chapter developed an informational framework for studying cybersecurity that brings together security and risk literatures, the philosophy of information, and new materialism. It substantiated three main assumptions that form the core of this framework: (1) cybersecurity is informational; (2) information is a peculiar entity; (3) cybersecurity should be theorised differently. This framework paves the way for investigating the ontology of information as the ultimate referent object of cybersecurity, and moving security logics away from existentiality, exceptionality, and emergency.

Firstly, cybersecurity is informational in essence. The connection between "cyber" and "information" in popular, academic, and policy discourses has roots in the evolution of cybernetics and information theory. Information is fundamental to computer science, software engineering, and all the sciences and technologies behind the tools and targets of cyber incidents. Computer science is even defined sometimes as the science of information-processing or information-transforming processes. All categories of information, be it syntactic, semantic, or pragmatic, are the core of cybersecurity threats and policies. Moving towards information thus marks a study of the ontology and materialities of cybersecurity as opposed to the conventional focus on discourses and speech acts in theorising the field.

Secondly, information is peculiar. Unlike new materialism that did not distinguish between the agential capacities of different types of matter or non-human things, the book argues that information is different. Information can be distinguished from matter and energy by its complexity, multiplicity, and transformational capacity. It is a fundamental force behind all objects' being and for driving many changes in the world. Specifically, in the next chapters, the book will focus on three main properties of information that co-produce the logics of cybersecurity. The first is intrinsic uncertainties of information systems; the second is the contingency and unpredictability of syntactic information (i.e., codes/software); and the third is the simultaneous physicality and non-physicality of information. Each of these three properties will be discussed in a separate chapter in relation to the logics of security in the infosphere.

Thirdly, given the informational ontology of cybersecurity and the peculiarities of information as such, it can be argued that cybersecurity is in turn peculiar. Understanding this peculiarity demands a new conceptual and methodological framework that deals with the anthropocentric limitations of security theories and attends to the fields' informational nature. In this framework, information is approached as a co-constitutive force of cybersecurity, whose influence is inherent to its existence and not necessarily bound to its relationship with other objects. Methodologically, the book focusses on information, rather than particular cyber incidents, as a co-productive, generative force in cybersecurity. In addition, discourse analysis is used in examining the empirical data by considering the intra-action between the material and the discursive together with the performativity of information. Using this non-anthropocentric informational framework, the book aims to move cybersecurity beyond the fixed logics of security and risk, towards the logics of negentropy, emergence, and noise, towards ultimately constructing the notion of entropic security.

In the following chapters, the book will show how the indeterminacies of information systems; the contingencies of codes/software; and the complex (non-)physicality of information engender peculiar logic(s) of security in cybersecurity, that differ profoundly from the aforementioned logics of security, and that are better studied through the notion of "entropic security." Entropic security, as an information-theoretic notion, moves cybersecurity from *absolute security to negentropy*, from *emergency to emergence*, and from *existentiality to noise*. This will be demonstrated by examining the disordered nature of the field of cybersecurity and its tendency towards increasing insecurity due to the intrinsic uncertainties of information systems. The book captures these uncertainties and disorders through the concept of entropy and reconceptualises cyber defence as "anti entropic practices" aiming at the avoidance of absolute insecurity; i.e., aiming at negentropy (Chapter 4). Afterwards, the book examines the contingencies and complexities of codes/ software and how they challenge human control through the logic of emergence. This is done by exploring how codes/software undermine the centrality of human intentionality as a basis for constructing enmity; and how they co-produce the subjects/objects of cybersecurity (Chapter 5). Finally, the (non-)physicality of information will be studied in light of how it reduces existentiality to being just *another* discourse in cybersecurity. This analysis will highlight the significance of *mundane* cybersecurity that can evoke urgency without existentiality, and therefore can be studied through the logic of noise (Chapter 6).

Notes

1 The assumption that information has to encompass meaning is opposed by some scholars who believe that information has an objective existence that does not depend on its perception, understanding, or interpretation. That is why, a distinction between "information" and "meaningful information" is sometimes made (Stonier, 2012).
2 The Macy conferences were interdisciplinary conferences held in the USA and are sometimes considered the most significant scientific events after the Second World War. Concepts like "information," "analogue/digital," and "feedback" were introduced in these conferences as part of regulatory frameworks that can apply to both humans and machines (Pias & Foerster, 2016).

References

Aalberts, T.E., & Werner, W.G. (2011). Mobilising Uncertainty and the Making of Responsible Sovereigns. *Review of International Studies, 37*(05), 2183–2200.

Aradau, C. (2004). Security and the Democratic Scene: Desecuritization and Emancipation. *Journal of International Relations and Development, 7*(4), 388–413.

Aradau, C. (2010). Security That Matters: Critical Infrastructure and Objects of Protection. *Security Dialogue, 41*(5), 491–514.

Aradau, C. et al. (2014). Discourse/Materiality. In C. Aradau, J. Huysmans, A. Neal, & N. Voelkner (Eds.), *Critical Security Methods: New Frameworks for Analysis* (pp. 57–84). Routledge.

Aradau, C., Lobo-Guerrero, L., & Van Munster, R. (2008). Security, Technologies of Risk, and the Political: Guest Editors' Introduction. *Security Dialogue, 39*(2–3), 147–154.

Aradau, C., & Van Munster, R. (2007). Governing Terrorism Through Risk: Taking Precautions,(un) Knowing the Future. *European Journal of International Relations, 13*(1), 89–115.

Aradau, C., & Van Munster, R. (2016). Poststructuralist Approaches to Security. In *Routledge Handbook of Security Studies* (pp. 75–84). Taylor and Francis.

Arquilla, J., & Ronfeldt, D. (1993). Cyberwar Is Coming! *Comparative Strategy, 12*(2), 141–165.

Arquilla, J., & Ronfeldt, D. (1996). Information, Power, and Grand Strategy: In Athena's Camp. *Significant Issues Series-Center for Strategic and International Studies, 18*, 132–180.

Ashby, W.R. (1958). *An Introduction to Cybernetics*. Chapman and Hall.

Backman, S. (2023). Risk Vs. Threat-Based Cybersecurity: The Case of the EU. *European Security, 32*(1), 85–103.

Balzacq, T. (2011). Enquiries into Methods: A New Framework for Securitization Analysis. In T. Balzacq (Ed.), *Securitization Theory: How Security Problems Emerge and Dissolve* (pp. 31–54). Routledge.

Balzacq, T., Leonard, S., & Depauw, S. (2015). The Political Limits of Desecuritization: Security, Arms Trade, and the EU's Economic Targets. In T. Balzacq (Ed.), *Contesting Security: Strategies and Logics* (pp. 104–121). Routledge.

Barad, K. (2003). Posthumanist Performativity: Toward an Understanding of How Matter Comes to Matter. *Journal of Women in Culture and Society, 28*(3), 801–831.

Barad, K. (2007). *Meeting the Universe Halfway: Quantum Physics and the Entanglement of Matter and Meaning*. Duke University Press.

Barbour, J. (2015). Bit from It. In A. Aguirre, B. Foster, & Z. Merali (Eds.), *It From Bit or Bit From It?: On Physics and Information* (pp. 197–212). Springer.

Beck, U. (1992). *Risk Society: Towards a New Modernity*. SAGE Publications Ltd.

Beck, U. (1999). *World Risk Society*. Polity.

Beck, U. (2002). The Terrorist Threat: World Risk Society Revisited. *Theory, Culture & Society, 19*(4), 39–55.

Beck, U. (2006). Living in the World Risk Society. *Economy and Society, 35*(3), 329–345.

Bellanova, R., Jacobsen, K.L., & Monsees, L. (2020). Taking the Trouble: Science, Technology and Security Studies. *Critical Studies on Security, 8*(2), 87–100.

Bendrath, R., Eriksson, J., & Giacomello, G. (2007). From 'Cyberterrorism' to 'Cyberwar', Back and Forth: How the United States Securitized Cyberspace. In J. Eriksson & G. Giacomello (Eds.), *International Relations and Security in the Digital Age* (pp. 57–82).

Bennett, J. (2009). *Vibrant Matter: A Political Ecology of Things*. Duke University Press.

Bennett, J. (2015). Systems and Things: On Vital Materialism and Object-Oritened Philosophy. In R.A. Grusin & R. Grusin (Eds.), *The Nonhuman Turn* (pp. 223–240). University of Minnesota Press.

Berry, D., & Dieter, M. (Eds.). (2015). *Postdigital Aesthetics: Art, Computation And Design*. Springer.

Bigo, D. (2000). When Two Become One: Internal and External Securitisations in Europe. In M. Kelstrup & M. Williams (Eds.), *International Relations Theory and The Politics of European Integration: Power, Security and Community* (pp. 171–204). Routledge.

Bigo, D. (2002). Security and Immigration: Toward a Critique of the Governmentality of Unease. *Alternatives*, *27*(1_suppl), 63–92.

Bigo, D. (2006). Security, Exception, Ban and Surveillance. In D. Lyon (Ed.), *Theorizing Surveillance: The Panopticon and Beyond* (pp. 46–68). Willan Publishing.

Bigo, D. (2008). Globalised (In)security: The Field aand the Ban-Opticon. In D. Bigo & A. Tsoukala (Eds.), *Terror, Insecurity and Liberty: Illiberal Practices of Liberal Regimes after 9/11* (pp. 5–49). Routledge.

Bishop, M. (2003). What Is Computer Security? *IEEE Security & Privacy*, *1*(1), 67–69.

Bogost, I. (2012). *Alien Phenomenology, Or, What It's Like to be a Thing*. U of Minnesota Press.

Booth, K. (1991). Security and Emancipation. *Review of International Studies*, *17*(4), 313–326.

Booth, K. (2007). *Theory Of World Security* (1st edition). Cambridge University Press.

Bourne, M., Johnson, H., & Lisle, D. (2015). Laboratizing the Border: The Production, Translation and Anticipation of Security Technologies. *Security Dialogue*, *46*(4), 307–325.

Bousquet, A. (2013). Welcome to the Machine: Rethinking Technology and Society through Assemblage Theory. In S. Curtis (Ed.), *Reassembling International Theory: Assemblage Thinking and International Relations* (pp. 91–97). Springer.

Braidotti, R. (2013). *The Posthuman*. John Wiley & Sons.

Braun, B., Schindler, S., & Wille, T. (2018). Rethinking Agency in International Relations: Performativity, Performances and Actor-Networks. *Journal of International Relations and Development*, *22*(4), 787–807.

Brunner, E.M., & Cavelty, M.D. (2009). The Formation of in-Formation by the US Military: Articulation and Enactment of Infomanic Threat Imaginaries on the Immaterial Battlefield of Perception. *Cambridge Review of International Affairs*, *22*(4), 629–646.

Bryant, L.R. (2011). *The Democracy of Objects*. Open Humanities Press.

Bryant, L.R. (2014). *Onto-Cartography: An Ontology of Machines and Media* (1st edition). Edinburgh University Press.

Buckland, M.K. (1991). *Information and Information Systems*. ABC-CLIO.

Burgin, M. (2010). *Theory of Information: Fundamentality, Diversity and Unification*. World Scientific.

Burgin, M., & Dodig-Crnkovic, G. (2013). *The Nature of Computation and the Development of Computational Models*. Computability in Europe 2013 (CiE 2013). https://pdfs.semanticscholar.org/3c18/1e61ef5a48cbcb084e68e93433ed40671612.pdf

Bynum, T.W. (2016). Informational Metaphysics: The Informational Nature of Reality. In L. Floridi (Ed.), *The Routledge Handbook of Philosophy of Information* (pp. 203–218). Routledge.

Ciută, F. (2009). Security and the Problem of Context: A Hermeneutical Critique of Securitisation Theory. *Review of International Studies*, *35*(02), 301.

Collins, A. (2005). Securitization, Frankenstein's Monster and Malaysian Education. *The Pacific Review*, *18*(4), 567–588.

Corry, O. (2012). Securitisation and 'Riskification': Second-Order Security and the Politics of Climate Change. *Millennium - Journal of International Studies*, *40*(2), 235–258.

Cristiano, F. et al. (Eds.). (2023). *Artificial Intelligence and International Conflict in Cyberspace*. Taylor & Francis.

Cudworth, E., Hobden, S., & Kavalski, E. (2018). Introduction: Framing the Posthuman Dialogues in International Relations. In E. Cudworth, S. Hobden, & E. Kavalski (Eds.), *Posthuman Dialogues in International Relations* (pp. 1–14). Routledge.

De Goede, M. (2004). Repoliticizing Financial Risk. *Economy and Society, 33*(2), 197–217.

Deacon, T.W. (2010). What is Missing from Theories of Information? In P. Davies & N.H. Gregersen (Eds.), *Information and the Nature of Reality: From Physics to Metaphysics* (pp. 146–169). Cambridge University Press.

Dillon, M. (2002). Network Society, Network-centric Warfare and the State of Emergency. *Theory, Culture & Society, 19*(4), 71–79.

Dlamini, M.T., Eloff, J.H.P., & Eloff, M.M. (2009). Information Security: The Moving Target. *Computers & Security, 28*(3–4), 189–198. https://doi.org/10.1016/j.cose.2008.11.007

Dourish, P. (2017). *The Stuff of Bits: An Essay on the Materialities of Information*. MIT Press.

Dourish, P., & Mazmanian, M. (2013). Media as Material: Information Representations as Material Foundations for Organizational Practice. In P.R. Carlile, D. Nicolini, A. Langley, & H. Tsoukas (Eds.), *How Matter Matters: Objects, Artifacts, and Materiality in Organization Studies* (pp. 92–118). OUP Oxford.

Dunn Cavelty, M. (2008a). *Cyber-Security and Threat Politics: US Efforts to Secure the Information Age*. Routledge.

Dunn Cavelty, M. (2008b). Cyber-Terror—Looming Threat or Phantom Menace? The Framing of the US Cyber-Threat Debate. *Journal of Information Technology & Politics, 4*(1), 19–36.

Dyson, F. (2001, March 13). *Is Life Analog or Digital?* https://www.edge.org/conversation/freeman_dyson-is-life-analog-or-digital

Elbe, S. (2008). Risking Lives: AIDS, Security and Three Concepts of Risk. *Security Dialogue, 39*(2–3), 177–198.

Eriksson, J. (2001). Cyberplagues, IT, and Security: Threat Politics in the Information Age. *Journal of Contingencies and Crisis Management, 9*(4), 200–210.

Floridi, L. (2009). Philosophical Conceptions of Information. In G. Sommaruga (Ed.), *Formal Theories of Information: From Shannon to Semantic Information Theory and General Concepts of Information* (pp. 13–53). Springer.

Floridi, L. (2013). *The Philosophy of Information*. OUP Oxford.

Floridi, L. (Ed.). (2016). *The Routledge Handbook of Philosophy of Information*. Routledge.

Floyd, R. (2016). Extraordinary or Ordinary Emergency Measures: What, and Who, Defines the 'success' of Securitization? *Cambridge Review of International Affairs, 29*(2), 677–694.

Fresco, N., & Wolf, M.J. (2016). Information Processing and Instructional Information. In L. Floridi (Ed.), *The Routledge Handbook of Philosophy of Information* (pp. 77–89). Routledge.

Friis, K., & Ringsmose, J. (Eds.). (2016). *Conflict in Cyber Space: Theoretical, Strategic and Legal Pespectives*. Routledge.

Gershenson, C. (2012). The World as Evolving Information. In A.A. Minai, D. Braha, & Y. Bar-Yam (Eds.), *Unifying Themes in Complex Systems VII: Proceedings of the Seventh International Conference on Complex Systems* (pp. 100–115). Springer Science & Business Media.

Gleick, J. (2011). *The Information: A History, a Theory, a Flood*. HarperCollins Publishers.

Grösser, S.N., & Zeier, R. (2012). *Systemic Management for Intelligent Organizations: Concepts, Models-Based Approaches and Applications*. Springer Science & Business Media.

Hagmann, J., & Dunn Cavelty, M. (2012). National Risk Registers: Security Scientism and the Propagation of Permanent Insecurity. *Security Dialogue, 43*(1), 79–96.

Hansen, L., & Nissenbaum, H. (2009). Digital Disaster, Cyber Security, and the Copenhagen School. *International Studies Quarterly, 53*(4), 1155–1175.

Harman, G. (2018). Object-Oriented Ontology: A New Theory of Everything. Penguin UK.

Harshman, N.L. (2016). Physics and Information. In L. Floridi (Ed.), *The Routledge Handbook of Philosophy of Information* (pp. 7–14). Routledge.

Hayles, N.K. (2008). *How We Became Posthuman: Virtual Bodies in Cybernetics, Literature, and Informatics*. University of Chicago Press.

Hoffmeyer, J. (2008). *A Legacy for Living Systems: Gregory Bateson as Precursor to Biosemiotics*. Springer Science & Business Media.

Hofkirchner, W. (2012). Emergent Information. When a Difference Makes a Difference…. *tripleC: Communication, Capitalism & Critique. Open Access Journal for a Global Sustainable Information Society, 11*(1), 6–12.

Hofkirchner, W. (2013). *Emergent Information: A Unified Theory of Information Framework*. World Scientific.

Jacobsen, K.L., & Monsees, L. (2019). Co-production: The Study of Productive Processes at the Level of Materiality and Discourse. In M. Hoijtink & M. Leese (Eds.), *Technology and Agency in International Relations* (pp. 24–41). Routledge.

Janich, P. (2018). *What Is Information?* University of Minnesota Press.

Jarvis, L., Macdonald, S., & Whiting, A. (2016). Analogy and Authority in Cyberterrorism Discourse: An Analysis of Global News Media Coverage. *Global Society, 30*(4), 605–623.

Kaltofen, C. (2018). With a Posthuman Touch: International Relations in Dialogue with the Posthuman—A Human Account. In E. Cudworth, S. Hobden, & E. Kavalski (Eds.), *Posthuman Dialogues in International Relations* (pp. 32–51). Routledge.

Kaminska, M. (2021). Restraint Under Conditions of Uncertainty: Why the United States Tolerates Cyberattacks. *Journal of Cybersecurity, 7*(1), tyab008.

Kessler, O., & Daase, C. (2008). From Insecurity to Uncertainty: Risk and the Paradox of Security Politics. *Alternatives, 33*(2), 211–232.

Knapp, K.J., Franklin Morris, R., Marshall, T.E., & Byrd, T.A. (2009). Information Security Policy: An Organizational-Level Process Model. *Computers & Security, 28*(7), 493–508.

Krahmann, E. (2011). Beck and Beyond: Selling Security in the World Risk Society. *Review of International Studies, 37*(01), 349–372.

Latour, B. (2005). *Reassembling the Social: An Introduction to Actor-Network-Theory*. OUP Oxford.

Law, J. (2002). Objects and Spaces. *Theory, Culture & Society, 19*(5–6), 91–105.

Law, J., & Singleton, V. (2005). Object Lessons. *Organization, 12*(3), 331–355.

Lawson, S. (2013). Beyond Cyber-Doom: Assessing the Limits of Hypothetical Scenarios in the Framing of Cyber-Threats. *Journal of Information Technology & Politics, 10*(1), 86–103.

Lawson, S.T. (2019). *Cybersecurity Discourse in the United States: Cyber-Doom Rhetoric and Beyond*. Routledge.

Leonardi, P.M. (2010). Digital Materiality? How Artifacts Without Matter, Matter. *First Monday, 15*(6–7). https://firstmonday.org/article/view/3036/2567

Lloyd, S. (2006). *Programming the Universe: A Quantum Computer Scientist Takes on the Cosmos*. Knopf Doubleday Publishing Group.

Lloyd, S. (2010). The Computational Universe. In P. Davies & N.H. Gregersen (Eds.), *Information and the Nature of Reality: From Physics to Metaphysics* (pp. 92–103). Cambridge University Press.

Lombardi, O. (2016). Mathematical Theory of Information (Shannon). In L. Floridi (Ed.), *The Routledge Handbook of Philosophy of Information* (pp. 30–36). Routledge.

Lundborg, T., & Vaughan-Williams, N. (2015). New Materialisms, Discourse Analysis, and International Relations: A Radical Intertextual Approach. *Review of International Studies*, *41*(1), 3–25.

Mahon, P. (2017). *Posthumanism: A Guide for the Perplexed*. Bloomsbury Academic.

McDonald, M., & Mitchell, A. (2017). Introduction: Posthuman International Relations. In C. Eroukhmanoff & M. Harker (Eds.), *Reflections on the Posthuman in International Relations: The Anthropocene, Security and Ecology*. E-International Relations.

McHarris, Wm.C. (2015). It from Bit from It from BIt…Nature and Nonlinear Logic. In A. Aguirre, B. Foster, & Z. Merali (Eds.), *It From Bit or Bit From It?: On Physics and Information* (pp. 225–234). Springer.

Mcmullin, E. (2010). From Matter to Materialism … And (almost) Back. In P. Davies & N.H. Gregersen (Eds.), *Information and the Nature of Reality: From Physics to Metaphysics* (pp. 13–37). Cambridge University Press.

Mitchell, A. (2014a). Only human? A worldly approach to security. *Security Dialogue*, *45*(1), 5–21.

Mitchell, A. (2014b, July 24). Dispatches from the Robot Wars; Or, What is Posthuman Security? *The Disorder Of Things*. https://thedisorderofthings.com/2014/07/24/dispatches-from-the-robot-wars-or-what-is-posthuman-security/

Ormes, E., & Herr, T. (2016). Understanding Information Assurance. In R. Harrison & T. Herr (Eds.), *Cyber Insecurity: Navigating the Perils of the Next Information Age* (pp. 3–18). Rowman & Littlefield Publishers.

Parks, L., & Starosielski, N. (2015). *Signal Traffic: Critical Studies of Media Infrastructures*. University of Illinois Press.

Pias, C. (2005). Analog, Digital, and the Cybernetic Illusion. *Kybernetes*, *34*(3/4), 543–550. https://doi.org/10.1108/03684920510581710

Pias, C., & Foerster, H.V. (2016). *Cybernetics: The Macy Conferences 1946–1953*. University of Chicago Press.

Piccinini, G., & Scarantino, A. (2016). Computation and Information. In L. Floridi (Ed.), *The Routledge Handbook of Philosophy of Information* (pp. 23–29). Routledge.

Primiero, G. (2016). Information in the Philosophy of Computer Science. In L. Floridi (Ed.), *The Routledge Handbook of Philosophy of Information* (pp. 90–106). Routledge.

Rasmussen, M.V. (2001). Reflexive Security: Nato and International Risk Society. *Millennium*, *30*(2), 285–309.

Rasmussen, M.V. (2004). It Sounds Like a Riddle': Security Studies, the War on Terror and Risk. *Millennium-Journal of International Studies*, *33*(2), 381–395.

Ratzan, L. (2004). *Understanding Information Systems: What They Do and why We Need Them*. American Library Association.

Salter, M.B. (2008). Imagining Numbers: Risk, Quantification, and Aviation Security. *Security Dialogue*, *39*(2–3), 243–266.

Schwarz, E. (2017). Hybridity and Humility: What of the Human in Posthuman Security? In C. Eroukhmanoff & M. Harker (Eds.), *Reflections on the Posthuman in International Relations: The Anthropocene, Security and Ecology* (pp. 29). E-International Relations.

Shannon, C.E. (1948). A Mathematical Theory of Communication. *The Bell System Technical Journal*, *27*, 379–423.

Sloman, A. (2011). What's Information, for an Organism or Intelligent Machine? How Can a Machine or Organism Mean? In G.D. Crnkovic (Ed.), *Information and Computation: Essays on Scientific and Philosophical Understanding of Foundations of Information and Computation*. World Scientific.

Solomon, T. (2015). Embodiment, Emotions, and Materialism in International Relations. In L. Åhäll & T. Gregory (Eds.), *Emotions, Politics and War* (pp. 58–70). Routledge.

Soni, J., & Goodman, R. (2017). *A Mind at Play: How Claude Shannon Invented the Information Age*. Simon and Schuster.

Steinhart, E. (1998). Digital Metaphysics. In T.W. Bynum & J.H. Moor (Eds.), *The Digital Phoenix: How Computers are Changing Philosophy* (pp. 117–134). Blackwell Publishers, Ltd.

Stevens, T. (2012). *Information Matters: Informational Conflict and the New Materialism*. Social Science Research Network. https://doi.org/10.2139/ssrn.2146565

Stevens, T. (2023). *What Is Cybersecurity For?* Policy Press.

Stockdale, L.P.D. (2013). Imagined Futures and Exceptional Presents: A Conceptual Critique of 'pre-Emptive Security'. *Global Change, Peace & Security*, 25(2), 141–157.

Stonier, T. (1991). Towards a new theory of information. *Journal of Information Science*, 17(5), 257–263.

Stonier, T. (2012). *Information and the Internal Structure of the Universe: An Exploration into Information Physics*. Springer Science & Business Media.

Stritzel, H. (2011). Security, the Translation. *Security Dialogue*, 42(4–5), 343–355.

The White House. (2023). *National Cybersecurity Strategy*. https://www.whitehouse.gov/wp-content/uploads/2023/03/National-Cybersecurity-Strategy-2023.pdf

Timpson, C. (2016). The Philosophy of Quantum Information. In L. Floridi (Ed.), *The Routledge Handbook of Philosophy of Information* (pp. 219–234). Routledge.

Trombetta, M.J. (2021). Security in the Anthropocene. In D. Chandler, F. Müller, & D. Rothe (Eds.), *International Relations in the Anthropocene: New Agendas, New Agencies and New Approaches* (pp. 155–172). Springer International Publishing. https://doi.org/10.1007/978-3-030-53014-3_9

Wheeler, J.A. (1992). Recent Thinking about the Nature of the Physical World: It from Bita. *Annals of the New York Academy of Sciences*, 655(1), 349–364.

Wiener, N. (1948). *Cybernetics: Or, Control and Communication in the Animal and the Machine*. Wiley & Sons.

Wilkinson, C. (2011). The Limits of Spoken Words: From Meta-narratives to Experiences of Security. In T. Balzacq (Ed.), *Securitization Theory: How Security Problems Emerge and Dissolve* (pp. 94–115). Routledge.

Williams, M.C. (2003). Words, Images, Enemies: Securitization and International Politics. *International Studies Quarterly*, 47(4), 511–531.

Williams, M.J. (2008). (In)Security Studies, Reflexive Modernization and the Risk Society. *Cooperation and Conflict*, 43(1), 57–79.

Wolfe, C. (2010). *What is Posthumanism?* U of Minnesota Press.

Zeilinger, A. (2005). *What do you believe is true even though you cannot prove it?* Edge.https://www.edge.org/response-detail/10380

Bousquet, A., Grove, J., & Shah, N. (2017). Becoming Weapon: An Opening Call to Arms. *Critical Studies on Security*, 5(1), 1–8.

Salter, M.B. (2015). Introduction: Circuits and Motions. In M. B. Salter (Ed.), *Making Things International 1: Circuits and Motion*. University of Minnesota Press.

4 Uncertainties and Disorders in Information Systems

Cyber Defence and the Logic of Negentropy

Introduction

Information, the previous chapter argued, is different from matter and energy. This difference entails a wide set of properties inherent to the existence of information which give it an autonomous status. Such properties are co-constitutive of peculiar logic(s) for cybersecurity that stand in tension with many theoretical assumptions about conventional security. This chapter focusses specifically on one property of the operation of information systems to illustrate this argument: intrinsic uncertainties and disorders. Studying uncertainties through an informational lens differs from existing understanding in the field of Security Studies. Although there is no independent literature on uncertainty as such, it has been extensively studied as part of theorising the concept and politics of *risk*. Uncertainty is sometimes viewed as a counter-concept to risk, assuming that the latter is linked to probability instead (Burgess, 2016; Petersen, 2016), or as integral to it (Aven, 2016). In other accounts, both risk and uncertainty are analysed as neo-liberal constructs (O'Malley, 2012), composed of multiple material and discursive elements, for governing what is projected as ungovernable (Aradau & Van Munster, 2007). Notwithstanding those different formulations, uncertainty has been largely approached as a point of distinction between security and risk; e.g., risk deals with the management of *uncertainty*, while security is concerned with the management of *insecurity* (Kessler & Daase, 2008).

This chapter, however, approaches the concept of uncertainty in a different way. It gives more weight in the analysis to uncertainty as such, as well as its temporalities and trajectories, than the ways it is tamed or governed by a human actor. Importantly, it presents an argument that uncertainty is an intrinsic property of the ontology of information and the operation of information systems. This property, in turn, co-produces a conceptualisation of cybersecurity that goes beyond the traditional distinctions between the logics of security and risk. This is a kind of security that the book conceptualises as "entropic security." Entropy is a concept that first originated in thermodynamics, but later moved to information theory and other academic fields, including economics, geography, and social theory. Although it is defined differently in

DOI: 10.4324/9781003454113-4

different sciences, entropy generally denotes uncertainty, disorder, noise, or disorganisation (Li and Du, 2017).

The chapter uses the definition of entropy as uncertainty and disorder both literally and analogically in presenting the logic of *negentropy* as the essence of cybersecurity and cyber defence. In a literal sense, entropy defined as uncertainty in information theory is used to investigate the indeterminacies of information systems and their implications on cybersecurity practices. It seeks to show the ontological nature of such uncertainties and their pervasiveness across the past, the present, and the future in cybersecurity. Analogically, entropy understood as disorder in cybernetics and thermodynamics is used to explore the peculiar temporalities, logic(s), and essence of cybersecurity. This informational analysis of uncertainties and disorders, the chapter argues, allows for a conceptualisation of cyber defence as *anti-entropic practices*; aiming at fighting the force of increasing disorder and absolute insecurity in cybersecurity; i.e., aiming at *negentropy* (negative entropy).

To unpack these arguments, the chapter starts with an explanation of the concept of entropy defined as uncertainty in information theory. It shows how entropy has always been intrinsic to the theorisation of information and its ontology on one side, and to the practical operation of information and communication systems on the other. The second section moves from theory to practice by analysing the uncertainties of the cybersecurity environment, which the chapter conceptualises as the "entropic space of cybersecurity." It focusses on four main examples of how this entropic space manifests itself: vulnerability analysis, intrusion detection, attribution and damage analysis, and lack of technical knowledge. In the third section, entropy is defined as disorder, based on cybernetics and thermodynamics. Here, entropy is used analogically to construct an understanding of the essence and logic(s) of cybersecurity as essentially entropic, i.e., tending towards disorder. It introduces the logic of negentropy as a way of understanding the essence of cybersecurity and cyber defence.

Entropy as uncertainty in information theory

Entropy first emerged in the 19th century as a major concept in the physical sciences, specifically in thermodynamics (Greven et al., 2014) – a field that studies energy and its transformations (Ness, 2012). Later, the concept moved to information theory in Claude Shannon's mathematical theory of communication (Shannon, 1948), and in Norbert Wiener's cybernetics (Wiener, 1948). Although entropy has been conceptualised and employed differently in these two fields, it can still be argued that "The Physicist's entropy and Shannon's entropy are two sides of a coin" (Siegfried, 2000, p. 66). Likewise, and as will be shown next, both formulations of the notion of entropy can provide important insights about the ontology of information and contribute to the theorisation of cybersecurity as entropic security.

Information has always been connected to the concept of entropy in the development of information theory. Shannon defined information as the measure of entropy in any system (Vedral, 2018). He assumed that when entropy (defined as uncertainty) increases, the amount of information also increases (Behrenshausen, 2016).[1] At first glance, this assumption that information is entropy (or uncertainty) may seem counter-intuitive. Understanding this claim requires an explanation of the probabilistic conceptualisation of information that Shannon introduced. In probability theory, information is defined by the surprise factor. If it is known that a certain event is highly probable, then getting informed about its occurrence is no surprise and no news, and thus no information. For example, being told that the sun will rise tomorrow does not provide any information because it is a well-known fact. Inversely, if something is perceived as improbable or unlikely, getting told about its occurrence represents a large amount of information (Lombardi, 2016). This can be seen as a relationship between probabilities and subjective degrees of beliefs that a particular event will occur, for instance (Milne, 2016). According to this understanding, Shannon presented information as inversely proportional to probability: the higher the probability (i.e., the less uncertainty), the less information, and vice-versa.

Applied to communication, Shannon assumed that when a particular message is sent by a source, this represents an activation of one possibility among many, because theoretically, all messages are likely to occur. Hence, there is an intrinsic level of uncertainty in communication systems linked to the choices that the sender makes about sending a message. If the message sent by the source is predetermined and known by the receiver, there would be no need for communication. Sending a specific message, thus, reduces the uncertainty linked to all the other messages that could have been sent but were not (Gregoire & Catherine, 2012). It is like tossing a coin: getting one side is one bit of information that reduces the uncertainty of two probable states. Here, information is not a given; it is rather "the progressive unfolding of this relation between uncertainty and certainty" (Malaspina 2018, p. 41). Put differently, when an event of a message unfolds, information works in a dynamic way, under various predictability and unpredictability levels.

In practical terms, the fact that information systems involve multiple layers of processing and interpretation leads to inevitable distortions. Distortions can result from overload, noise, truncation, inconsistency, imprecisions, etc. Accordingly, any information representation can be seen as essentially "controlled distortions of the original information" (Ratzan, 2004). Furthermore, the interaction of information systems with thermodynamic subsystems results in further propagation of uncertainty and growing entropy. For instance, in communication devices, information transmits as energy from one end to the other, i.e., pulses of light or waves. It is inevitable that during the transmission process some energy gets distorted, subject to disturbances in the channel through which that information is being transmitted. This means that information that reaches the receiver can never be identical to the original.

Uncertainty is intensified in the case of digital information systems as compared to analogue ones. As argued by one study, "The essence of the bit is the uncertainty inherent in it" (Kafri & Kafri, 2013, p. 135). Analogue means of communication are less sensitive to noise due to "data redundancy." For example, in the case of an analogue transmission of a picture, the image received is an exact version of the original one. If the analogue transmission process is subjected to noise, e.g., blur, the quality of the received image may be reduced, but it will not be totally destroyed. So, if one or more pixels are corrupted, it will just leave a blank spot, while the rest of the picture remains intact. But in case of digital transmission, the existence of noise can corrupt the entire picture. For example, the colour blue in a digital picture showing a blue sky is encoded by a number of bits that if distorted will distort the entire picture's colour (Kafri & Kafri, 2013). Because computing involves a huge number of operations, it is prone to error, and any such error will lead to inevitable loss of information (Keyes, 1977).

The underlying idea behind this analysis is that entropy is a default state. Any system that does not have uncertainty is a system without information (Behrenshausen, 2016). Decreasing such entropy is one definition of communication: when communication occurs, it is capable of reducing entropy (Cannizzaro, 2016). Furthermore, communication channels are always noisy, and *minimising* – rather than entirely eliminating – such noise is key for successful information transmission. In these systems, information involves the act of selection among various probabilities to reduce uncertainty, which is in itself a process marked by lots of indeterminacies (Cannizzaro, 2016). Thus, although uncertainty is not a property of information as such, it can be said that it is a property of its existence. Put differently, information in itself does not engender uncertainty, yet uncertainty is intrinsic to its conceptualisation and to its ontological makeup. This is perfectly put by one scholar who argues "…information is carved from that entropic space" (Wicken, 1987, pp. 180–181).

The entropic space of cybersecurity

The theoretical link between information and entropy, as well as its manifestation in the operation of information and communication systems, similarly unfolds in cybersecurity. Cybersecurity too is carved from an entropic space, where uncertainties are ontological and mark a default state. These uncertainties have far-reaching implications on the sort of policies implemented to secure against cyber threats and importantly, on the span of the *possible* in cybersecurity. The entropic space of cybersecurity redefines the temporality of uncertainty as theorised by some security-risk literature. The uncertainties of cybersecurity are not necessarily future-oriented or concerned with future unknowns (for example: Stockdale 2015); they are also linked to the past and the present. To illustrate this point, four examples will be discussed to show some facets of the entropic space of cybersecurity, in which different types of uncertainties and temporalities come together.

Vulnerability analysis

All networks, processes, operating systems, applications, and even devices in information systems operate through lines of codes written by programmers and software vendors. A security vulnerability is an error or a bug in coding, design, or modelling that can be exploited for attacking the system or network. Bugs are either introduced intentionally for malicious use, often called "backdoors," or unintentionally as part of unplanned implementation or design errors. They vary in their severity and exploitability; not all bugs are exploitable and thus not all necessarily qualify as *vulnerabilities*. Some may give the attackers limited privileges on the system and others may allow for full remote control of the compromised target (Ablon & Bogart, 2017).

Unknown vulnerabilities are sometimes called "zero-day vulnerabilities" or "zero-days" for short. They are the ones that are unknown to the software vendors and for which no patch (fix) is available. Hence the name "zero-day," which refers to the number of days the vulnerability was known to the target before it is exploited. These are the most difficult to detect and thus difficult to defend against. Because of that, a large market for zero-days is becoming very popular, in which governments and several other entities, and even individuals, participate.[2] These markets are primarily "information markets," since they sell knowledge about the potential results of exploiting specific vulnerabilities – knowledge that the vendor/user/target does not possess. As said by a zero-day vulnerability seller and noted by one study "we don't sell weapons, we sell information" (Fidler, 2016, p. 280).

The majority of reported cyber incidents result from the exploitation of bugs (errors) in software coding. Any single software contains millions of coding lines, in which errors are mostly inevitable. Consequently, it is widely acknowledged by most cybersecurity actors that all software would always have bugs of some sort. Some bugs can stop the system from working, interrupt connectivity, or cause irregularities in the operation of certain devices. Others, however, are security vulnerabilities that can be exploited for hostile cyber operations that affect the confidentiality, integrity, and availability of information (Winkler & Gomes, 2016). According to one study, on average, programmers make 15–50 errors in every 1000 lines of coding; a number that may slightly decrease after testing (Hage, 2020). This inevitability of bugs has always been emphasised by almost all software programmers and engineers. They acknowledge that "there will always be another bug" in every software, considering the complexity of information systems. It is even argued by some experts that errors are more prevalent and more problematic in software design than in any other technology, and that no matter how skilful the people performing software reviews are, some bugs will remain undiscovered. This creates a belief that bugs and the uncertainties they produce are "endemic to programming" and that they are the "natural hazards" of information systems (Nissenbaum, 1997).

Inevitable vulnerabilities mean inevitable uncertainties and a large number of unknowns and *unknowables*. It is almost impossible to know beforehand how a certain software would react with the hardware it is installed on or with other programs on the system (Ormes & Herr, 2016). Furthermore, even in the case of known vulnerabilities for which patches are available, there is no guarantee that a patch will solve the problem. In many cases, a patch can contain more security holes that are not discoverable unless applied or tested on the system. The unpredictability of how the system will react to the applied patch and how the elements of that system will interact with it adds another level of uncertainty that challenges human control (Libicki et al., 2015).

Intrusion detection

Intrusion detection systems (IDSs) refer to processes through which malicious activities on the system are detected by analysing the possible signs of incidents that violate the system's security policy and standard practices. Intrusion prevention systems (IPSs) involve a similar process, but one that does not stop at detection. IPSs seek to prevent the intrusion from succeeding or spreading. It can do so by stopping the intrusion itself and terminating the connection used in the attack, and thus blocking access to the target, or by changing the environment of the system configuration and removing/changing the malicious content of the attack (Scarfone & Mell, 2010).[3]

Unlike the military sector in which the first steps of an attack may be detectable, the non-physical elements of information – which will be discussed further in Chapter 6 – can make cyber intrusions invisible to the target for a long period of time. One report that investigated attack behaviour in enterprise production environments showed that security alerts are generated for only 9% of recorded cyber incidents and that 53% of attacks infiltrate unnoticed (Mandiant, 2022). Additionally, data shows that the average number of days to identify breach incidents was 207 and 70 days to contain them in 2022 (IBM Security, 2023); rising from 197 in 2018 and 191 in 2017 (IBM Security, 2019). It is also often other entities – particularly the FBI in the US case – that notify the victims that their systems have been compromised (*Protecting America from Cyber Attacks* 2015, p. 8).

These challenges are strongly linked to the inherent multiplicity of information and its emergent properties. Since complex information systems are diverse and constantly changing, it is challenging to keep a static image of the topology of networks. Given this complexity, the data produced by IDSs is often ambiguous. It is common for attacks to be mistaken for errors or over-load signals and vice-versa. This is because the existence of anomalies is a fundamental characteristic of certain systems (Lazarevic et al., 2005). In consequence, based on a survey conducted in 2022 on 813 security professionals across ten industries in five countries, 43% of respondents said that 40% of their security alerts are false positives: indicating a threat that does not exist (Orca Security, 2022). This happens despite the continuous progress in advanced IDSs.

Attribution and damage analysis

Another major source of uncertainty in cybersecurity is the problem of attribution. Attribution refers to the process of identifying the source of the attack, the identity of the attacker, and their incentives. In fact, there are many technical barriers related to the nature of the internet as a packet-switching network that impedes attribution and forensic analysis. Firstly, cyber incidents can be initiated by botnets, or compromised devices that are centrally controlled by the attacker or "the botnet operator." This means that the attack is distributed among different nodes of thousands of machines with multiple IP addresses, and thus making it difficult to correlate the malicious packets to a single source.

Secondly, the use of proxies is another source of complication because it changes packets' IP addresses and provides anonymity over the internet for privacy purposes or otherwise. For example, big corporations sometimes use Network Address Translation (NAT), a technology that reduces the number of IP addresses shown publicly. The use of this technology could result in linking any potential intrusion with the institution itself rather than a certain node inside it. Onion routing is another example of a method used for maintaining communication anonymity by hiding the IP address of the sender. The onion router operates through a layered cryptographic system that encrypts the packet and its content from a hop (i.e., onion router) to another, keeping the sender anonymous. Thirdly, ISPs use dynamic IP addresses, meaning that a new IP address will be assigned to the user every time they connect to the internet. This means that the packets a certain user sends and receives may have different IP addresses over time (Boebert, 2010).

Fourthly, there is the problem of what Buchanan refer to as "the dilemma of interpretation" or knowing the real intentions of the intruder in this environment of uncertainty (Buchanan, 2016). In the traditional military sector, an invasion of troops is often interpreted as an offensive attack and a breach of sovereignty even if the intruder's intention was enhancing their defence. Yet, in cybersecurity, a system intrusion can be done for the mere purpose of intelligence gathering, even among friendly nations, without damaging the system or resulting in any kind of harm. This leaves the targeted state with a dilemma of whether to interpret this as an intelligence collection, contingency planning, or as a preparation for a cyber-attack. This is what Buchanan called the security dilemma of cybersecurity that is arguably more complicated than the security dilemma of conventional sectors (Buchanan, 2016). This does not mean that attribution is not technically possible, it rather shows the severity of the uncertainties that overshadow attribution as one part of cybersecurity's threat logics and policy response. This is also important considering that sometimes even the main target of the attack is not known in the very early stages, particularly if the attack spreads across a large number of systems (Egloff, 2020).

Another level of uncertainty can be seen in the process of damage analysis following a cyber incident. Even after detection, and notwithstanding attribution, it is difficult to determine with great certainty the full extent of the resulting damage. To date, there is no agreed-upon protocol for defining and measuring cyber damage; most of the available figures are inaccurate estimates. Consequently, if multiple companies were targeted by the same attack, each might have its own different estimates of the damage scale and resulting losses. This all makes it very difficult to predict an attack on a particular target in the future, or to anticipate the full magnitude of the resulting damage or costs. Since the data available is not enough, quantifying cyber risk to make generalisable statistics remains very problematic. As argued by Steven Bellovin, a researcher on computer network and security who also worked as technical leader and fellow at AT&T laboratory, in a congressional hearing: "A locksmith can tell you how long a safe can resist an attack with certain kinds of tools. A computer scientist can't do the same" (*Cybersecurity – Getting It Right*, 2003, p. 18).

Lack of awareness and technical knowledge

It is not just the absence of information that creates uncertainties in the infosphere; it is also the absence of *technical knowledge* and *awareness* even when information is available, which is sometimes referred to in cybersecurity discourses as the "knowledge gap." This aspect can be framed within the subjective property of information mentioned in the previous chapter. Even when information is available, it might not be informative for everyone. In cybersecurity, the informative quality of information is bound by the technical nature and complexity of information systems. Given the complexity and indeterminacy of such systems, understanding many cybersecurity problems requires a certain level of awareness and expertise that may not be available outside certain circles of cybersecurity experts. This is one explanation why the majority of cyberattacks target known vulnerabilities for which patches are already available. For example, in a survey of over 3000 security experts in nine countries, including the USA, respondents indicated that 60% of system breaches in 2019 were linked to known vulnerabilities for which patches were available (Ponemon Institute, 2019, p. 5). In fact, many known vulnerabilities go unpatched because the users or operators of the system are either unaware of them, unaware of the patching methods, or struggling to keep up with the great number of patches that are issued monthly. It is common that users ignore patches or run outdated software even in cases of critical vulnerabilities; not knowing the possible cascading effects this may have on the security of everyone (Backman, 2023). Hence, discovering vulnerabilities and issuing patches is never the end of the road in cybersecurity.

Again, the nondeterministic nature of information systems is one factor that complicates the patching process. Patching is an error-prone process and one that is never fully controllable by humans. It is difficult to predict how the system's elements are going to react to the applied patches. In some cases, applying a patch may result in even more security holes or lead to disabling the system altogether. This problem intensifies in the case of Industrial Control Systems (ICS) through which critical information infrastructure is operated. On the ICS vendors side, there is reluctance to issue patches straight away following the knowledge of certain vulnerabilities, because multiple tests and validations have to be performed first to see how the software operation would react to the patch. On the users' side, there is a need to do similar testing for the patch on the ICS-specific environment to minimise the chance of unintended consequences. This is a time-consuming process, especially that the system would need to be taken down completely for the tests to be done and the patches to be applied. That is why, patching a system, particularly ICS, may happen over a long period of time – sometimes years – after the vulnerability is discovered (Lee, 2016). It can be thus argued that the more complex and vital the system is, the more challenging the process of patching will be. According to Steven Bellovin:

> …the composition of systems, the components of complex systems working together properly is a very, very difficult and unsolved problem…It is not that the administrators are irresponsible, or that the vendors haven't supplied good tools, it is that we don't know how to do it easily, reliably and without breaking something else.
>
> (*Cybersecurity – Getting It Right*, 2003, pp. 43, 44)

Although it may be in private companies' interest, like AT&T that Bellovin represented in this congressional hearing, to evade cybersecurity liability by presenting patching as a particularly complex task, there are many reasons to believe that it often is. Furthermore, technical knowledge about patching and system configuration is usually lacking in the case of ordinary users, who are the most targeted by hostile cyber incidents. The fact that users sometimes cannot differentiate between secure and unsecure software diminishes their power to influence the market of cybersecurity or to push for more secure software by-design. This indirectly encourages the "fix it later" culture that drives software vendors to introduce buggy software in the market, and then issue incremental patches gradually; meanwhile subjecting software to possible exploitation of potential vulnerabilities (Chong, 2016). This knowledge gap does not only exist between ordinary and skilled users, but also between local and federal governments on one side, and the public and private sector on the other side. A federal executive once noted in a congressional hearing in 2005 that when they were communicating patching information with town supervisors, one replied saying: "I don't understand what you mean by patching. When I hear the word, I look for duct tape" (The Future of Cyber and Telecommunications Security at DHS, 2006, p. 60).

These are just some of the many examples of the protracted entropic space of cybersecurity, co-produced by the peculiarities of information. It is a space in which uncertainty cannot be simply reduced to a human discourse, empirical non-knowledge, or to a particular future temporality. It is *ontological uncertainty* that is intrinsic to the information systems that cybersecurity aims to protect and that is pervasive across the past, the present, and the future. This is not a transformation from security to risk that follows Ulrich Beck's argument on the emergence of a "risk society" in the second modernity, in which the unknown, incalculable, and uncontrollable dangers are massively increasing (Beck, 1992). It is also not a transition from the semantic field of security to that of risk as suggested by risk studies that discuss modern conceptualisation of security (Kessler & Daase, 2008). It is a recognition that cybersecurity is *born* a risk society as such due to its informational ontology. This is a security environment that the concept of entropy – defined as uncertainty – directly captures. Meanwhile, there is a lot more that entropy can add to the theorisation of cybersecurity, its evolutionary processes, and its essence. This can be made clearer by exploring another meaning for entropy as defined in physics: entropy as disorder.

Entropy as disorder and the essence of cybersecurity

Another way entropy appeared in the information theory literature was in Wiener's cybernetics (Wiener, 1948, 1988). Wiener assumed that there is a natural connection between information and entropy by presenting information as the measure of the system's organisation and entropy as the measure of its *disorganisation*. Thus, unlike Shannon who viewed information as entropy, information to Wiener is the inverse of entropy, or negative entropy. This is an idea that Wiener derived from physics and the second law of thermodynamics. In physics, entropy represents the assumption that matter and energy are always changing, and in every change they lose part of their structure and of their information too. Any physical change of matter-energy produces a certain loss of order or information, and entropy is the measure of that lost information. This is part of a metaphysical view of the universe that sees time as moving in one direction: towards more loss of information and order, which defines the "ultimate fate of every physical entity" (Bynum, 2008, p. 17).

Physical entropy explains the natural tendency of ice to melt, for example, and for hot things to eventually cool and lose their heat, unless acted upon by an external force to reverse that (e.g., putting the melted ice back in a fridge). Even when entropy seems to be decreasing in parts of a system, it must be increasing in others. So, for instance, if a fridge is cooling and experiencing a decline in entropy, it is because heat is coming out of it and consequently raising the entropy of its surrounding. Thereby, it is assumed that "the entropy of the universe never goes down" (Davies, 2019, p. 32). From a thermodynamics perspective, information processing of all types, be it encoding, transmission, decoding, etc., leads to energy dispersion and increases entropy (Li and Du, 2017). This is how many scholarly contributions connect information theory and physics (Davies, 2019), as in the work of Rolf William Landauer (Landauer, 1991, 1999).

Entropy as a security analogy

Entropy as disorder can serve as an informative analogy to think about cybersecurity. *Firstly*, the analogy of entropy can help in understanding why some cyber insecurities are not just accepted, but often also embraced as natural product of complex information systems. When Wiener conceptualised information as negative entropy (i.e., the inverse of entropy), he did not mean that entropy itself is negative. To him, entropy demonstrates reality's natural gravitation towards disorganisation and chaos (Wiener 1948, 1988). Similarly, the mathematician Warren Weaver distinguished between what he called "spurious uncertainty," resulting from the influence of noise that could disrupt information communication, and "desirable uncertainty" that reflects the sender's "freedom of choice" in relation to deciding what messages to send (Weaver 1949). Even when viewed negatively as a problem of communication, noise remains integral to the existence of information. As argued by Malaspina, "…the creation of information can only occur on the basis of noise" (Malaspina 2018, 75).

In cybersecurity, it is widely believed among many actors that "complexity is the worst enemy of security" and that "everything about complexity leads towards lower security" (*Overview of the Cyber Problem*, 2003, p. 11). The complex operation of information systems creates an understanding of absolute or complete security as something that contradicts the very essence of information being dynamic and complex. In a congressional hearing following the targeting of the computer systems of the US Bureau of Industry and Security by hackers based in China, a cybersecurity strategist said: "…I don't believe you can ever have a dynamic, effective, productive system and be 100 percent secure. It would violate the reason why you built it" (*Cyber Insecurity*, 2007, p. 55). Again, here, complexity – and in turn insecurity – is ontological. It does not simply reflect a human perception or choice of "embracing risks" for commercial or economic gains as part of risk governance (Amoore, 2013).

Ontological uncertainties that are normalised, or even embraced, in cybersecurity can be explained by the computer scientist Frederick P. Brooks' distinction between "accidental" and "essential" complexity in software engineering – in line with Weaver's distinction between spurious and desirable uncertainty as explained above. For Brooks, the complexity of software systems and digital computers is incomparable to any other artefact, because of the large numbers of forms they can have and the non-linear interactions among their elements. These complexities are not accidental; they are *essential* to the existence of software. Though many methods have been developed to handle accidental complexities, there is "no silver bullet" for dealing with essential complexities (Brooks, 1987). Likewise, as argued by an adviser in a risk and insurance company explaining the growing importance of cyber insurance, "…we discovered that there is no silver bullet and that there is always going to be some residual risk, despite how strong your practices are" (*The Role of Cyber Insurance in Risk Management*, 2016, p. 37). Again, even though cyber insurance companies would understandably advocate a belief that cyber insecurity is inevitable, as stated previously, it is a widely shared belief by most actors. In this way, cybersecurity becomes inherently a kind of entropic security, based on an understanding that cyber threats are intrinsic to the very makeup of the systems that need protection.

Secondly, the analogy of entropy adds an important temporal dimension to the conceptualisation of security. Thermodynamics entropy is connected to a particular view of progress that distinguishes the past from the future, referred to as "the arrow of time." It is an assumption that certain physical phenomena are irreversible. Melting ice cubes are an example of a spontaneous phenomenon for which reversibility – i.e., water turning back into ice – does not happen naturally and requires an external source of energy, e.g., a fridge (Kisak, 2015). Time, according to this view, points at the direction of increasing entropy, and hence, increasing disorder (Zuchowski, 2024). This temporal aspect of entropy and its arrow of time is evident in cybersecurity too. Many actors believe that cyber dependency and the complexity of information systems are unavoidable, increasing, and mostly irreversible. Many discourses frame the cyber threat around the idea that cyber dependence "will continue to increase" (The Department of Defense, 2015, p. 9), and that complexity cannot be changed. For example, data shows that in the period between 2014 and 2019, the likelihood of a data breach grew by 31%; malicious and criminal attacks increased by 21%; and the average total cost of a data breach increased by 12% (IBM Security, 2019). Similarly, in 2023, the average cost of data breach reached an all-time high $4.45 million, which is an increase of 15.3% from 2020 (IBM Security, 2023). Accordingly, the future of cybersecurity is conceived as essentially more threatening than the present. Just like the one-directional physical entropy, cyber threats are seen as always evolving, ever-increasing, and becoming more serious every day. As expressed by a representative from the DHS describing this cybersecurity challenge: "the train has left the station" (Cybersecurity, 2010, p. 40).

It follows that being "secure" in cybersecurity is perceived by most actors as more of a *process* than a *goal* or a *state* per se. Secure is the application of enough security measures that cope with the ever-increasing cyber threats, which do not necessarily eliminate them. That being the case, normative adjectives are often attached to cybersecurity, such as "good security," "weak security," "agile security," "bad security," etc., which all seem counter-intuitive to the semantically positive notion of security. Security in this way becomes a non-binary concept that breaks the distinctions between "secure" and "insecure." This processual, temporal nature of entropic security is best exemplified in statements like "cybersecurity is a journey and not a destination" that is repeated by the government (*Protecting Cyberspace*, 2011, p. 21) and the private sector alike (*Securing America's Future*, 2012, p. 43). As Tim Stevens argues, "…no one seriously believes that we can engineer digital networks in a way that will make cyber insecurity vanish" (Stevens, 2023, p. 36). That is, entropic security is a process, not an end goal or a status, because absolute security is unachievable.

Thirdly, entropy is additive, and therefore is a signification of complex interdependencies. The entropy of a system is calculated by summing up the entropy of each of its parts. This implies an intrinsic connection between all parts of the system and its overall thermodynamic properties. An air conditioner that is decreasing the entropy of a room by cooling it down is actually dispersing hot air that increases the overall environment's entropy (Kisak, 2015). The same logic applies to cybersecurity. As entropic security, cybersecurity is constituted by every single user of digital technologies, from individual citizens to corporations and governments. This creates an understanding that "cyber security is only as strong as its weakest link" (*Protecting Cyberspace as a National Asset*, 2010, p. 17), since an attack against one may affect the security of the rest. As one study put it, "Everybody is your neighbor on the Internet" and calculating one's risk has to consider that of others (Aycock, 2006, p. 3). Due to this interdependency, end-users are sometimes blamed for not updating their systems regularly, not necessarily for the sake of their own security, but for the negative implications of that on the security of the state. This is acknowledged by the most recent US National Cybersecurity Strategy which considered the challenge of placing "too much" responsibility on individual users and small organisations as a "systemic challenge" in cybersecurity (The White House, 2023).

By extension, even if entropy appears to be decreasing in one part of the system, it is increasing in others, as explained earlier in the fridge and air conditioner examples. This idea captures a very complex paradox in cybersecurity. Although in other security fields, such as economic or military security, more development can be linked to increasing security, this is not the case in cybersecurity. Arguably, cyberspace is becoming less secure even as security technologies improve. There is a general belief that the more cyber dependent the state is, the more inevitably threatened it becomes. As argued by Stephen E. Flynn in 2012, then the co-director of George J Kostas Research Institute for Homeland Security at Northeastern University, "the United States is at risk of becoming a victim of its own success," given its leading position in using information technology systems and applications (*America Is Under Cyber Attack*, 2012, p. 39). This was also reiterated in the National Cybersecurity Strategy of 2023: "The world is entering a new phase of deepening digital dependencies" (The White House, 2023, p. 2). This growing insecurity does not just contradict with the development of security technologies, but also with the application of more security measures and more investment in cybersecurity. It is estimated that the global cybersecurity market has grown to $173 billion in 2020 from $88 billion in 2012, despite the simultaneous exponential rise in cyber threats (Columbus, 2020). This entropic nature is already acknowledged by the majority of cybersecurity actors and scholarly literature, even if they do not use the term "entropy" as such.

From absolute security to negentropy

Thus far, the chapter has made the case for the relevance of the concept of entropy for understanding the intrinsic uncertainties of cybersecurity and its disorderly nature. But how can entropy describe processes in the opposite direction; ones that aim at increasing order and organisation? Assuming that cybersecurity is entropic does not deny the existence of a wide range of cybersecurity defence measures that target cyber threats/risks. It also does not dismiss the possibility of achieving progress in cybersecurity. But such processes have their own peculiarities, and there remain many useful insights that the analogy of entropy can add in this respect.

In general terms, the assumption that the entropy of the universe is always increasing may seem to contradict humans' and modern sciences' strive for more order and organisation (Arnheim, 2010). However, many physicists would argue that the existence of entropy and increasing disorder does not negate the possibility of *negentropy*: increase in *order* and *organisation*, i.e., negative entropy. While the entropy of the entire universe keeps increasing, some local systems, such as the fridge or air conditioner in the previously mentioned examples, may experience increasing order or negentropy. As put by Wiener: "There are local and temporary islands of decreasing entropy in a world in which the entropy as a whole tends to increase, and the existence of these islands enables some of us to assert the existence of progress" (Wiener, 1988, p. 36). This means that individual parts of the system can experience negentropy (decreasing entropy) even if the entropy of the entire system is rising.

This argument can be broadened to apply, not just to parts of a system, but also to the universe as a whole. From a cosmological point of view, on one side, some scholars predict that the entropy of the universe will keep increasing until it reaches the maximum point of heat death. They often link this to technology too by arguing that the increasing use of technology will inevitably lead to increasing the entropy of the universe. This is framed as "the diminishing returns of technology" or the assumption that rather than solving our problems, technological developments decrease the overall amount of available energy in the world. Such argument has been applied to other issues too, such as industrial capitalism and its modes of production that lead to overpopulation, pollution, and other negative implications that contribute to the demise of energy in the universe (Best, 1991; Rifkin & Howard, 1989). On the other hand, nonetheless, this view is challenged by other scholars who contend that the continuing expansion of the world, through evolution, biological reproduction, and processes of acquiring knowledge, show a lot of negentropy and therefore can reverse entropy (Sonne, 1985 and Kisak, 2015).

This debate is partially connected to a more general one on the relationship between order and chaos. Many scholars believe that chaos is not necessarily the opposite of order. They talk about the possibility of order and regularities emerging out of chaos and irregularities; i.e., "irregular regularities" (Best, 1991, p. 203). This is what Toffler argues in the foreword of a book named *Order out of Chaos*, speaking about how entropy can be the creator of order, even in a spontaneous manner (Prigogine & Stengers, 2018). In the same vein, instead of viewing uncertainty antagonistically, it is sometimes approached positively as a possible driver for creativity and freedom, particularly in certain fields like arts and academic research (Smithson, 2012). According to Wiener, machines, just like humans, can exert such "anti-entropic processes" (Wiener, 1988, p. 32). Similarly, Floridi assumes that informational entities are capable of decreasing the entropy of the universe, or at least resisting it

(Floridi, 2013). Some scholars even talk about what they call "extropy," which is introduced as a post-humanist or even trans-humanist idea that technology will change human existence in an opposite way to the chaos that entropy suggests (Pepperell, 1995).

Negentropy that can increase in local parts of the system, despite the increasing entropy of the whole, is analogous to cybersecurity. Even though the overall cybersecurity statistics all over the world show rising insecurity every year, there are always areas of improvement in specific fields, industries, organisations, or periods of time. As an example, though the total global average cost of system breaches has been increasing over the years, the year 2017 and 2019 witnessed a considerable drop in costs (IBM Security, 2019, 2023). Likewise, though the overall breaches are increasing in number, a system glitch as a root cause for those breaches decreased between 2015 and 2017 (IBM Security, 2019) and decreased slightly again in 2020 (IBM Security, 2020). The inevitability of cyber insecurity makes such numbers look good in security evaluation even if the overall picture is getting worse. This is because, as argued by Chandler, failure is intrinsic to complex, non-linear systems – ones that informational systems resemble. "Failing better," thus, becomes a more achievable target than achieving complete success (Chandler, 2014). This also reflects the nature information processes that, as argued by Malaspina, represents a "controlled way of falling," "recuperated disorganisations," or "repeated cycles of acquisition and loss of equilibrium" (Malaspina 2018, 73).

Hence, instead of defence strategies, cybersecurity measures are better theorised as *anti-entropic practices*. Non-action, or not taking any measures to secure "cyberspace," results in increasing insecurity because entropy is the default state. Cybersecurity interventions, thus, seek to resist entropy in individual systems, even when the overall cyber insecurity is rising. That is, cyber defence measures as anti-entropic practices are processes aiming at the avoidance of absolute cyber insecurity rather than achieving absolute security. Framing cybersecurity measures as anti-entropic aiming at negentropy recognises the fact that such measures are not always directed towards a particular threat or a constitutive causality, but against the entropic force of increasing insecurity.

Given this, cybersecurity as entropic security explains the dominant discourse adopted by many actors and scholarly literature that the essence of cybersecurity is the reduction of risks, threats, and vulnerabilities rather than their elimination. Such belief is not exclusive to the private sector as it might seem. As an example, it was explicitly mentioned in the presidential policy directive of 2013 on cybersecurity that: "The terms 'secure' and 'security' refer to *reducing* [emphasis added] the risk to critical infrastructure by physical means or defense cyber measures to intrusions, attacks, or the effects of natural or manmade disasters" (The White House, 2013). A cybersecurity policy document issued by the DHS also defined the state of critical infrastructure's security by that in which the owners and operators of such systems "manage risks" and sustain "adequate security" (Department of Homeland Security, 2011, p. 11). Furthermore, similar to what Wiener described as "temporary islands of decreasing entropy" (Wiener, 1988, p. 36), the state of security in cybersecurity can be as temporary. The multiplicity and transformational capacity of information results in an understanding of cybersecurity as a "moving target" or "a snapshot of a moment in time" (*Cyber Insecurity*, 2007, p. 55), given its speed of change and dynamism.

The world is constantly evolving, accompanied by rapidly changing threat environment. This change is seen as the only constant in information environments since the development of the first internet, ARPANET, making it difficult to catch up with every next vulnerability or threat. For example, a system that is updated and patched can still have hundreds of unknown vulnerabilities, and a detected malware can change its signature making its first discovery ineffective. Therefore, patching is not entirely controllable by humans. A system that is updated today and cleaned of all vulnerabilities can have hundreds of vulnerabilities that the user did not know about a day after. Even when a single component of an information system is presumably secure, there is no guarantee that it will remain as such when interacting with other components in an information system or infrastructure.

Hence, cybersecurity as entropic security is measured by the relative future improvements to the conditions of the present and the past. It is not totally fixed to the logics of defence against a present and urgent threat as some security theories would suggest, but rather aims at making systems relatively more secure in the future. This creates an understanding of the essence of security in cybersecurity as the management of risks and reduction of uncertainties. Entropic security, thus, does not entirely subscribe to traditional logics used to distinguish between both risk and security in the literature. This can be made clearer by investigating the kind of policies and practices put forward to achieve cybersecurity in such an entropic space.

Anti-entropic practices in the infosphere

The entropy of information systems imposes conditions that enables, shapes, and/or limits the kind of discourses that actors construct in cybersecurity, and the sort of policies, strategies, and practices introduced to manage cyber risks/ threats. All strategies of intervention proposed in cybersecurity assume a high level of uncertainty. Additionally, as shown previously, the uncertainties of cybersecurity are not limited to the future. Uncertainties are intrinsic in thinking about both the present and past threats. Hence, the traditional distinction between uncertainty as incomplete, incalculable knowledge and risk as measurable uncertainties, as argued in some security studies literature, does not necessarily hold in cybersecurity. Uncertainty lies at the centre of every single strategy applied to achieve cybersecurity (Scala et al., 2019); something that is evident in both the practice and discursive levels.

Before a cyber incident takes place, several long-term precautionary measures are usually proposed in order to reduce the potential threat/hazard and its damaging consequences. Although some risk literature assumes that precaution implies the possibility of prevention (Aradau, 2016), this possibility is perceived differently in cybersecurity. Cybersecurity as entropic security implies that cyber threats are generally continuous. Intrusions happen in massive numbers, and not all intrusions have the same level of criticality. There is a general agreement on the need for risk prioritisation, which means that prevention would be possible for only few incidents, but not all. Prevention appears more when the object of inquiry is CNIs, which many argue should be approached with "zero-tolerance."

Importantly, prevention in cybersecurity as entropic security is not necessarily about stopping one big incident, crises, or disaster. It is mainly about the prevention of cascading effects, attack spread, or stopping as much incidents as possible so that the overall threat is reduced, but not necessarily eliminated. Again, making such practices anti-entropic rather than strictly defensive and aiming at negentropy rather than absolute security. Consequently, given this entropic nature of cybersecurity and the fact that most incidents happen with little to no warning, response is considered the core of cybersecurity policies and strategies. If most attacks cannot be predicted or prevented, the best that can be done is responding to them, by stopping them, limiting their damaging implications, and recovering from their consequences.

Vulnerability reduction

As stated earlier, vulnerabilities are seen as by-products of complexity in information systems. The more complex systems are, the faster they change, the more vulnerabilities they produce, referred to as "complexity induced vulnerability" or "negative technological synergy" (Hellström, 2007). Therefore, the proposed policies/strategies to deal with vulnerabilities usually aim at their assessment, reduction, mitigation, and management, but not necessarily their elimination or total prevention. To that end, proposed and implemented practices include manufacturing secure software ("secure by design" or "secure out of the box") by automating coding, improving training for code developers, and implementing robust quality assurance that involves external certification and third-party reviews. They also include measures to ensure secure software deployment by developing easy-to-deploy, use, and configure products.

Yet, just like physical negentropy that can only succeed in individual systems, there is a high level of risk acceptance and prioritisation in choosing the targets of cybersecurity measures. As an example, many software vendors postpone security measures for after the software is released by issuing patches, instead of spending more time creating secure software from the start. Some companies launch bug bounty programmes later to encourage researchers and white-hat hackers to discover vulnerabilities in their systems before they get exploited so that they fix them. This can be described as a postponed precautionary measure or as explained by chief technology officer in McAffee "It would be like taking a decongestant or a pain reliever when you have a cold, rather than eating healthy and exercising and building your immunity" (*America is Under Cyber Attack: Why Urgent Action Is Needed*, 2012, p. 31).

Modelling, simulations, and exercises

Another anti-entropic method to manage the uncertainty and unpredictability of the infosphere, and to prepare for a cyber-attack, is the performance of modelling, simulations, and exercises. These methods are as old as the 1970s when penetration-testing was used to ensure the functionality of the system throughout the process of software and hardware development. They can be considered a form of "anticipatory security" to "inhabit the future," implying that vulnerability and failure are inevitable. Such method is particularly important in cybersecurity since a lot of the threat scenarios, especially the ones associated with major destructions, are either few or non-existent (Stevens, 2015). These practices imply the impossibility of prevention, and thus aim at managing the future threats by simulating them for better response when an attack takes place. It operates under a worst-case-scenario assumption; i.e., preparing for the worst because it cannot be prevented.

Prevention through access control

Access control is inherently an anti-entropic practice rather than one that simply defends against threats. It refers to a number of methods used to prevent unauthorised access to certain information systems, and to ensure that only authorised users can use them. It is commonly stated as part of the preparedness strategy to prevent, limit, and later identify cyber-attacks. It is done through authentication by granting access to particular users; using unique identifiers (usernames, biometrics, etc.) to identify those users; validating the users' identity; and finally logging files and records to associate actions to particular users for analysis (Kim & Solomon, 2016). Besides, there is always continuous calls for improving encryption and the use of VPNs, firewalls, and antivirus software. However, the multiplicity of information and the huge number of diverse nodes in every system that resembles entropy's microstates limit the utility of such measures. The more microstates, the higher the entropy and vice-versa. As stated by Bellovin in Congress:

> We are moving more towards a motel rather than a hotel model. In the hotel, there are one or two entrances and everyone is walking past the front desk. In the motel, every room has got its own door to the outside. It is a lot harder to secure that, and we are moving more towards that ladder.
>
> (*Cybersecurity – Getting It Right*, 2003, p. 35)

Deterrence and pre-emption

Although deterrence seems like an imposed analogy on cybersecurity that does not really work, it is still sometimes used as part of cybersecurity discourses, particularly in that of the military. Many studies have argued against the applicability of deterrence to cybersecurity, particularly given the uncertainties of attribution (Lindsay, 2015). However, Joseph Nye stated four main ways through which deterrence and dissuasion can take place in cybersecurity. The first is deterrence by punishment, or the use of retaliatory measures that do not have to be restricted to cyber. This is the type of deterrence in which the uncertainties of attribution play a major role. This is because an attack needs to be attributed to a certain source for punishment to be possible. The second is deterrence by denial, by improving cyber defences and building resilience. The third is entanglement, by investing in more interdependencies that raise the cost of the attack. Finally, there is deterrence by norms, or the "reputational costs" that can be imposed on an attacker in a way that harms their soft power (Nye, 2017). Stevens also discussed this normative aspect of deterrence as a way to change states' behaviour in cyberspace and deter adversaries, away from the traditional conceptualisation of military power (Stevens, 2012).

Among these four methods, the first two are the dominant ones in cyber deterrence discourses. Deterrence by punishment is especially mentioned when discussing cyber crimes and the fact that the absence of strong punishments does not provide adequate deterrence to stop cyber criminals. Strong defence and resilience are also portrayed as critical steps towards deterring the enemy. Yet, added to these two, offensive cyber operations are sometimes proposed as a necessary intervention to dissuade attackers, based on perceptions like "the best defence is a good offense." In practice, it is now common that states conduct operations by exploiting vulnerabilities in the target system and then characterising them as defence operations, if direct damage was not inflicted. The DoD in its cyber strategy of 2015 mentioned that it may perform cyber operations for the sake of deterrence and defeating adversaries (The Department of Defense, 2015). This is sometimes framed as "active cyber defence," which may include non-disruptive practices like maintaining a presence on the adversary's network for information collection, or disruptive ones like "hacking back" to recover stolen data for instance (Healey, 2019; Pattison, 2020). However, pre-emptive measures in cybersecurity are still bound by its chronic uncertainties. This is because vulnerabilities may be patched and thus hindering the attempts to exploit them; exploits may result in unintended consequences; or a targeted system may prove resilient to exploitation (Valeriano et al., 2018).

Intrusion detection and prevention

IDSs and IPSs take place under a high level of uncertainty. As explained earlier, it is impossible for such processes to provide a completely accurate image about a certain system. In addition, if they are not "rapid" and "timely," detection and prevention's usefulness will be limited. This is again another example for how prevention has completely different logics in cybersecurity. The fact that intrusion prevention has to be preceded by detection means that the intrusion or at least its first steps have to occur first for them to be stopped or prevented. Hence, prevention inherently aims at stopping an intrusion that has already taken place in order to limit its progression or its damaging consequences, rather than preventing its occurrence. A successful prevention is therefore one that *quickly* detects an intrusion and *mitigates* its harm, not necessarily one that stops the intrusion from happening in the first place.

Resilience

In the risk literature, some scholars deal with resilience as a response strategy to disaster or crises (Krieger, 2016), while others associate it with the epistemic regime of novelty and surprises (Aradau, 2014). It is sometimes defined as the ability of the system to "bounce back" to a "normal" status quo that needs to be preserved (Krieger, 2016; Petersen, 2016). In cybersecurity, resilience is an overarching concept that spans various logics and temporalities. It is proposed as both part of the preparedness and response strategies. It is the ability of the system to overcome intrusions or minimise the scale of their consequences, as per measures taken *before* an attack takes place. It is the survivability and adaptability of the system *during* an attack and its capacity to function properly and survive on redundancy. It is the system's ability to recover the consequences of the attack *after* it occurs. Moreover, due to the entropic nature of cybersecurity, resilience is not only fixed to disasters or crises and is not limited to novelty and surprises. Besides, given the parallels of the arrow of time and irreversibility in physical entropy, resilience as an anti-entropic process does not imply a "normal state" that needs to be preserved, because the cyber threat is perceived as continuous in which the "normal" does not really exist. That is, the present is not fully secure, so bouncing back to it is not a target per se; rather, it is negentropy that is achievable.

Conclusion

This chapter focussed on the definition of entropy as uncertainty and disorder to outline the first logic that governs cybersecurity as entropic security, which is the logic of negentropy (negative entropy). On one hand, it harnessed insights from information theory's understanding of entropy to theorise the intrinsic uncertainties of cybersecurity. On the other hand, physical entropy – defined as disorder in cybernetics and thermodynamics – was used as an analogy to describe the disordered nature of cybersecurity and add a temporal aspect to the analysis. The result is a theorisation of cybersecurity as entropic security that is carved out of an entropic space through anti-entropic processes aiming at negentropy.

Entropy has always been an integral part of information theory and of Shannon's theory of communication. Defined as uncertainty, the chapter demonstrated how entropy is the default state that communication in essence tries to minimise. Entropy as a security analogy has the capacity to capture the peculiarities of uncertainties and disorders in cybersecurity in multiple ways. Firstly, it shows how such uncertainties are ontological, and therefore cannot be reduced to an empirical challenge of "not knowing." Such uncertainties exist regardless of human perception and even with the existence of knowledge about threats. Because of their ontological nature, uncertainties in cybersecurity are also *not necessarily* future-oriented; they span the past, the present, and the future. Secondly, and by extension, entropy is a signification of uncertainties that are essential to the operation of information systems, and hence are partially embraced as a natural product of complexity. Thirdly, the uncertainties and disorders that entropy represents have a particular temporal nature that adds a processual conceptualisation to cybersecurity. Lastly, entropy reflects a theorisation of uncertainty that is connected to complex interdependencies among information systems. As shown in the chapter, such an understanding of uncertainties and disorders is already acknowledged by most cybersecurity actors – even without coining the term "entropy" as such. In this way, entropic security adds an inductive conceptual framework to study existing understandings of the field that cannot be easily captured by the paradigms of security and risk.

Due to the complexity of information systems, and the big number of coding lines required for the operation of any software, bugs are mostly inevitable. Even when vulnerability analysis is effective, there will always be another unknown bug that could be exploited in a cyber incident. Moreover, applying a patch may result in more security holes or to system malfunction, and it is usually difficult to predict the result. In addition, intrusions may stay invisible to the target for a long time and IDSs may be overwhelmed with false negatives and false positives. Related to this is the problem of attributing attacks to a certain source and accurately measuring the damage inflicted. The use of botnets, proxies, NAT, onion routing, and dynamic IP addresses are examples of factors that pose technical barriers to attribution. Finally, the technical nature of information systems requires a certain level of awareness or expertise to understand their security issues and fix them properly. This level may not be available to ordinary, non-expert users: the largest pool of targets.

Operating in such an entropic space, cybersecurity becomes a disordered field analogous to entropy in the field of thermodynamics. Just like the natural tendency of matter-energy to decay, cybersecurity has a tendency towards increasing disorder, due to the complexity and dynamism of information systems. For that reason, the proportionality of increasing complexity and increasing insecurity is evident in many cybersecurity discourses. Besides, a similar arrow of time to that of entropy can be found in thinking about cyber dependency and increasing future threats; both are seen as irreversible. Entropic security is thus less about the state of being secure and more about fighting the entropic forces of insecurity. Additionally, entropic security is additive, in which the insecurity of the part contributes to the insecurity of the whole. Yet, paradoxically, the same as physical entropy, collective cyber insecurity may be increasing even if the security of individual systems is improving and if the state of the technology is progressing.

Accordingly, in cybersecurity, defence is reflected in anti-entropic processes that aim at achieving negentropy. Negentropy in cybersecurity is achieved in two ways. The first is decreasing the entropy of particular systems. Since securing everything is not possible, a risk-based prioritisation of the targets that receive resources and attention should be adopted. The second is shifting the point of absolute cyber insecurity further away, rather than seeking absolute cyber security. This negentropy has an inherent temporary nature, since cybersecurity is a moving target and information systems are essentially dynamic. Non-action would mean insecurity because entropy is a default state, unless external force is applied. External forces or anti-entropic practices in cybersecurity include: vulnerability reduction, modelling and simulations, deterrence and pre-emption, intrusion detection and prevention, and resilience. They all reflect an understanding that even though prevention is important, what is more feasible is the quick detection of intrusions, mitigation of the resulting damage, and resilience in the form of the survivability of the targeted vital systems.

Notes

1 Some scholars regarded this assumption about the direct proportionality between information and entropy and both concepts being synonyms as a logical contradiction in Shannon's argument. For more information, see (Stonier, 2012).
2 The markets of vulnerabilities and zero-day exploits can be classified into: white, grey, and black markets. White markets are usually the bug bounty programmes run by many companies such as Google, Facebook, and Microsoft that bug hunters participate in with the responsibility to disclose found bugs to those companies. Grey markets are the ones that the intelligence agencies and governments participate in for either offensive or defensive purposes. Black markets are used solely by criminals for malicious use (Libicki et al., 2015).
3 A very important example for detection and prevention systems is the EINSTEIN program run by the DHS. EINSTEIN is a system that aims at detecting and blocking attacks at the federal level and providing information to the DHS for accurate situational awareness for public and private partners. The first EINSTEIN was launched in 2003 with the aim of recording the network traffic in all federal civilian agencies in the executive branch for it to be analysed and for malicious activities to be identified. EINSTEIN 2 used signature-based methods to issue alerts on potential attacks. EINSTEIN 3 accelerated (E3A) was developed in 2012 not just for detection but for actively stopping the detected attacks (*EINSTEIN*, n.d.).

References

Ablon, L., & Bogart, A. (2017). *Zero Days, Thousands of Nights: The Life and Times of Zero-Day Vulnerabilities and Their Exploits*. Rand Corporation.
America is Under Cyber Attack. (2012). *Why Urgent Action is Needed: Hearing before the Subcommittee on Oversight, Investigation, and Management, of the Committee on Homeland Security* (Serial No. 112-85), U.S. House of Representatives, 112th Cong.
Amoore, L. (2013). *The Politics of Possibility: Risk and Security Beyond Probability*. Duke University Press.
Aradau, C. (2014). The Promise of Security: Resilience, Surprise and Epistemic Politics. *Resilience, 2*(2), 73–87.

Aradau, C. (2016). Risk, (in)security and International Politics. In A. Burgess, J.O. Zinn, & A. Alemanno (Eds.), *Routledge Handbook of Risk Studies.* Routledge, Taylor & Francis Group.

Aradau, C., & Van Munster, R. (2007). Governing Terrorism Through Risk: Taking Precautions,(un) Knowing the Future. *European Journal of International Relations, 13*(1), 89–115.

Arnheim, R. (2010). *Entropy and Art: An Essay on Disorder and Order.* Univ of California Press.

Aven, T. (2016). The Reconceptualization of Risk. In A. Burgess, J.O. Zinn, & A. Alemanno (Eds.), *Routledge Handbook of Risk Studies* (pp. 58–72). Routledge, Taylor & Francis Group.

Aycock, J. (2006). *Computer Viruses and Malware.* Springer.

Backman, S. (2023). Normal Cyber Accidents. *Journal of Cyber Policy, 8*(1), 114–130.

Beck, U. (1992). *Risk Society: Towards a New Modernity.* SAGE Publications Ltd.

Behrenshausen, B.G. (2016). *Information in Formation: Power and Agency in Contemporary Informatic Assemblages* [Ph.D., The University of North Carolina at Chapel Hill]. https://search.proquest.com/docview/1805474652/abstract/FD12EAFA BFF54374PQ/1

Best, S. (1991). Chaos and Entropy: Metaphors in Postmodern Science and Social Theory. *Science as Culture, 2*(2), 188–226.

Boebert, W.E. (2010). A Survey of Challenges in Attribution. *Proceedings of a Workshop on Deterring Cyberattacks: Informing Strategies and Developing Options for U.S. Policy,* 41–52.

Brooks, F.P. (1987). No Silver Bullet Essence and Accidents of Software Engineering. *Computer, 20*(4), 10–19.

Buchanan, B. (2016). *The Cybersecurity Dilemma: Hacking, Trust and Fear Between Nations.* C. Hurst & Co Publishers.

Burgess, A. (2016). Introduction. In A. Burgess, A. Alemanno, & J.O. Zinn (Eds.), *Routledge Handbook of Risk Studies.* Routledge, Taylor & Francis Group.

Bynum, T.W. (2008). Norbert Wiener and the Rise of Information Ethics. In J. Weckert & J. Van Den Hoven (Eds.), *Information Technology and Moral Philosophy* (pp. 1–25). Cambridge University Press.

Cannizzaro, S. (2016). The Philosophy of Semiotic Information. In L. Floridi (Ed.), *The Routledge Handbook of Philosophy of Information* (pp. 290–303). Routledge.

Chandler, D. (2014). *Resilience: The Governance of Complexity.* Routledge.

Chong, J. (2016). Bad Code: Exploring Liability in Software Development. In R. Harrison, T. Herr, & R.J. Danzig (Eds.), *Cyber Insecurity: Navigating the Perils of the Next Information Age* (pp. 69–86). Rowman & Littlefield Publishers.

Columbus, L. (2020, April 5). 2020 Roundup Of Cybersecurity Forecasts And Market Estimates. *Forbes.* https://www.forbes.com/sites/louiscolumbus/2020/04/05/2020-roundup-of-cybersecurity-forecasts-and-market-estimates/

Cybersecurity. (2003). *Getting It Right: Hearing of the subcommittee on Cybersecurity, Science, and Research, and Development, before the Select Committee on Homeland Security* (Serial No. 108-18), U.S. House of Representatives, 108th Cong.

Cyber Insecurity. (2007). *Hackers are Penetrating Federal Systems and Critical Infrastructure: Hearing before the Subcommittee on Emerging Threats, Cybersecurity and Science and Technology, of the Committee on Homeland Security House of Representatives* (Serial No. 110-26), U.S. House of Representatives, 112th Cong.

Cybersecurity: *DHS' Role, Federal Efforts, and National Policy: Hearing before the Committee on Homeland Security* (Serial No. 117-71), U.S. House of Representatives, 111th Cong. (2010).

Davies, P. (2019). *The Demon in the Machine: How Hidden Webs of Information Are Finally Solving the Mystery of Life.* Penguin UK.

Department of Homeland Security. (2011). *Blueprint for a Secure Cyber Future*. https://www.dhs.gov/xlibrary/assets/nppd/blueprint-for-a-secure-cyber-future.pdf

Egloff, F.J. (2020). Public attribution of cyber intrusions. *Journal of Cybersecurity*, 6(1), tyaa012.

EINSTEIN. (n.d.). Department of Homeland Security. Retrieved 15 October 2018, from https://www.dhs.gov/einstein

Fidler, M. (2016). Government Acquisition and Use of Zero-Day Software Vulnerabilities. In R. Harrison & T. Herr (Eds.), *Cyber Insecurity: Navigating the Perils of the Next Information Age* (pp. 3–18). Rowman & Littlefield Publishers.

Floridi, L. (2013). *The Ethics of Information*. OUP Oxford.

Gregoire, N., & Catherine, N. (2012). *Foundations Of Complex Systems: Emergence, Information And Prediction* (2nd Edition). World Scientific.

Greven, A., Keller, G., & Warnecke, G. (Eds.). (2014). *Entropy*. Princeton University Press.

Hage, J. (2020). Solved and Open Problems in Type Error Diagnosis. *CEUR Workshop Proceedings*, 2707, 62–74.

Healey, J. (2019). The Implications of Persistent (and Permanent) Engagement in Cyberspace. *Journal of Cybersecurity*, 5(1), tyz008.

Hellström, T. (2007). Critical Infrastructure and Systemic Vulnerability: Towards a Planning Framework. *Safety Science*, 45(3), 415–430.

IBM Security. (2019). *Cost of a Data Breach Report* (p. 76). Ponemon Institute. https://www.ibm.com/downloads/cas/ZBZLY7KL

IBM Security. (2020). *Cost of a Data Breach Report*. https://www.ibm.com/security/digital-assets/cost-data-breach-report/

IBM Security. (2023). *Cost of a Data Breach Report*. https://www.ibm.com/reports/data-breach?utm_content=SRCWW&p1=Search&p4=43700075239448391&p5=p&gclid=CjwKCAjw8symBhAqEiwAaTA__GUGrx15AETYUWIyWTlVltCTNt5ad1U6FwE7QJNjuD2_5WB12-j6KRoCMakQAvD_BwE&gclsrc=aw.ds

Kafri, O., & Kafri, H. (2013). *Entropy: God's Dice Game*. Createspace Independent Pub.

Kessler, O., & Daase, C. (2008). From Insecurity to Uncertainty: Risk and the Paradox of Security Politics. *Alternatives*, 33(2), 211–232.

Keyes, R.W. (1977). Physical Uncertainty and Information. *IEEE Transactions on Computers*, C–26(10), 1017–1025.

Kim, D., & Solomon, M.G. (2016). *Fundamentals of Information Systems Security*. Jones & Bartlett Publishers.

Kisak, P.F. (Ed.). (2015). *Entropy and Negentropy: The End and the Beginning*. Createspace Independent Publishing Platform.

Krieger, K. (2016). Resilience and Risk Studies. In A. Burgess, J.O. Zinn, & A. Alemanno (Eds.), *Routledge Handbook of Risk Studies* (pp. 335–343). Routledge, Taylor & Francis Group.

Landauer, R. (1991). Information is Physical. *Physics Today*, 44(5), 23–29.

Landauer, R. (1999). Information Is a Physical Entity. *Physica A: Statistical Mechanics and Its Applications*, 263(1–4), 63–67.

Lazarevic, A., Kumar, V., & Srivastava, J. (2005). Intrusion Detection: A Survey. In V. Kumar, J. Srivastava, & A. Lazarevic (Eds.), *Managing Cyber Threats: Issues, Approaches, and Challenges* (pp. 19–78). Springer US.

Lee, R.M. (2016). Protecting Industrial Control Systems in Critical Infrastructure. In R. Harrison, T. Herr, & R.J. Danzig (Eds.), *Cyber Insecurity: Navigating the Perils of the Next Information Age* (pp. 31–46). Rowman & Littlefield Publishers.

Li, D., & Du, Y. (2017). *Artificial Intelligence with Uncertainty*. CRC Press.

Libicki, M.C., Ablon, L., & Webb, T. (2015). *The Defender's Dilemma: Charting a Course Toward Cybersecurity*. Rand Corporation.

Lindsay, J.R. (2015). Tipping the Scales: The Attribution Problem and the Feasibility of Deterrence Against Cyberattack. *Journal of Cybersecurity*, 1(1), 53–67.

Lombardi, O. (2016). Mathematical Theory of Information (Shannon). In L. Floridi (Ed.), *The Routledge Handbook of Philosophy of Information* (pp. 30–36). Routledge.

Malaspina, C. (2018). *An Epistemology of Noise.* Bloomsbury Publishing.

Mandiant. (2022). *Deep Dive into Cyber Reality Security Effectiveness Report 2020 Report.* https://mandiant.widen.net/s/gsvtgb5hdj/security-effectiveness-report-2020

Milne, P. (2016). Probability and Information. In L. Floridi (Ed.), *The Routledge Handbook of Philosophy of Information* (pp. 15–22). Routledge.

Ness, H.C.V. (2012). *Understanding Thermodynamics.* Courier Corporation.

Nissenbaum, H. (1997). Accountability in a Computarized Society. In B. Friedman (Ed.), *Human Values and the Design of Computer Technology* (pp. 41–64). Cambridge University Press.

Nye, J.S. (2017). Deterrence and Dissuasion in Cyberspace. *International Security, 41*(3), 44–71.

O'Malley, P. (2012). *Risk, Uncertainty and Government.* Routledge-Cavendish.

Orca Security. (2022). *Cloud Security Alert Fatigue Report.* https://orca.security/wp-content/uploads/2022/03/Orca-2022-Cloud-Security-Alert-Fatigue-Report.pdf

Ormes, E., & Herr, T. (2016). Understanding Information Assurance. In R. Harrison & T. Herr (Eds.), *Cyber Insecurity: Navigating the Perils of the Next Information Age* (pp. 3–18). Rowman & Littlefield Publishers.

Overview of the Cyber Problem. (2003). *A Nation Dependent and Dealing with Risk, Hearing of the Subcommittee on Cybersecurity, Science, and Research, and Development, before the Select Committee on Homeland Security* (Serial No. 108-13), U.S. House of Representatives, 108th Cong.

Pattison, J. (2020). From defence to offence: The ethics of private cybersecurity. *European Journal of International Security, 5*(2), 233–254.

Pepperell, R. (1995). *The Post-Human Condition.* Intellect Books.

Petersen, K.L. (2016). Risk and Security. In M. Dunn Cavelty & T. Balzacq (Eds.), *Routledge Handbook of Security Studies* (pp. 117–125). Routledge, Taylor & Francis Group.

Ponemon Institute. (2019). *Costs and Consequences of Gaps in Vulnerability Response.* ServiceNow. https://www.servicenow.com/content/dam/servicenow-assets/public/en-us/doc-type/resource-center/analyst-report/ponemon-state-of-vulnerability-response.pdf

Prigogine, I., & Stengers, I. (2018). *Order Out of Chaos: Man's New Dialogue with Nature.* Verso.

Protecting America from Cyber Attacks. (2015). *The Importance of Information Sharing: Hearing before the Committee on Homeland Security and Governmental Affairs* (Serial No. 114-412), U.S. Senate, 114th Congress.

Protecting Cyberspace as a National Asset. (2010). *Comprehensive Legislation for the 21st Century: Hearing before the Committee on Homeland Security and Governmental Affairs* (Serial No. 111-1103), U.S. Senate, 111th Cong.

Protecting Cyberspace. (2011). *Assessing the White House Proposal, Hearing before the Committee on Homeland Security and Governmental Affairs* (S. Hrg. 112-221), U.S. Senate, 112th Cong.

Ratzan, L. (2004). *Understanding Information Systems: What They Do and why We Need Them.* American Library Association.

Rifkin, J., & Howard, T. (1989). *Entropy: Into the Greenhouse World.* Bantam Books.

Scala, N.M., Reilly, A.C., Goethals, P.L., & Cukier, M. (2019). Risk and the Five Hard Problems of Cybersecurity. *Risk Analysis, 39*(10), 2119–2126.

Scarfone, K., & Mell, P. (2010). Intrusion Detection and Prevention Systems. In P. Stavroulakis & M. Stamp (Eds.), *Handbook of Information and Communication Security* (pp. 177–192). Springer Science & Business Media.

Securing America's Future. (2012). *The Cybersecurity Act of 2012, Hearing before the Committee on Homeland Security and Governmental Affairs*, U.S. Senate (S. Hrg. 112-524), Cong. 112th.

Shannon, C.E. (1948). A Mathematical Theory of Communication. *The Bell System Technical Journal, 27*, 379–423.

Siegfried, T. (2000). *The Bit and the Pendulum: From Quantum Computing to M Theory—The New Physics of Information*. WILEY.

Smithson, M. (2012). The Many Faces and Masks of Uncertainty. In G. Bammer & M. Smithson (Eds.), *Uncertainty and Risk: Multidisciplinary Perspectives* (pp. 13–26). Routledge.

Sonne, J.C. (1985). Entropic Communication in Families with Adolescents. *International Journal of Family Therapy*, 7(3), 178–191.

Stevens, T. (2012). Information Matters: Informational Conflict and the New Materialism (SSRN Scholarly Paper ID 2146565). Social Science Research Network. https://doi.org/10.2139/ssrn.2146565

Stevens, T. (2015). *Cyber Security and the Politics of Time* (1st edition). Cambridge University Press.

Stevens, T. (2023). *What Is Cybersecurity For?* Policy Press.

Stockdale, L.P.D. (2015). *Taming an Uncertain Future: Temporality, Sovereignty, and the Politics of Anticipatory Governance*. Rowman & Littlefield.

Stonier, T. (2012). *Information and the Internal Structure of the Universe: An Exploration into Information Physics*. Springer Science & Business Media.

The Department of Defense. (2015). *The Department of Defense Cyber Strategy*. http://www.dtic.mil/doctrine/doctrine/other/dod_cyber_2015.pdf

The Future of Cyber and Telecommunications Security at DHS. (2006). *Hearing before the Subcommittee on Economic Security, Infrastructure Protection, and Cybersecurity, of the Committee on Homeland Security* (Serial No. 109-102), U.S. House of Representatives, 109th Cong.

The Role of Cyber Insurance in Risk Management. (2016).*Hearing before the Subcommittee on Cybersecurity, Infrastructure Protection, and Security Technologies, of the Committee on Homeland Security* (Serial No. 114-61), U.S. House of Representatives, 114th Cong.

The White House. (2013). *Presidential Policy Directive—Critical Infrastructure Security and Resilience*. https://obamawhitehouse.archives.gov/the-press-office/2013/02/12/presidential-policy-directive-critical-infrastructure-security-and-resil

The White House. (2023). *National Cybersecurity Strategy*. https://www.whitehouse.gov/wp-content/uploads/2023/03/National-Cybersecurity-Strategy-2023.pdf

Valeriano, B., Jensen, B., & Maness, R.C. (2018). *Cyber Strategy: The Evolving Character of Power and Coercion*. Oxford University Press.

Vedral, V. (2018). *Decoding Reality: The Universe as Quantum Information*. Oxford University Press.

Weaver, W. (1949). The Mathematics of Communication. *Scientific American*, *181*(1), 11–15. JSTOR.

Wicken, J.S. (1987). Entropy and Information: Suggestions for Common Language. *Philosophy of Science*, *54*(2), 176–193. https://doi.org/10.1086/289369

Wiener, N. (1948). *Cybernetics: Or, Control and Communication in the Animal and the Machine*. Wiley & Sons.

Wiener, N. (1988). *The Human Use Of Human Beings: Cybernetics And Society*. Hachette UK.

Winkler, I., & Gomes, A.T. (2016). *Advanced Persistent Security: A Cyberwarfare Approach to Implementing Adaptive Enterprise Protection, Detection, and Reaction Strategies*. Syngress.

Zuchowski, L. (2024). *From Randomness and Entropy to the Arrow of Time*. Cambridge University Press.

5 Informational Contingencies

From Emergency to Emergence

Introduction

For many years, it has been widely acknowledged by various actors that "Cyberspace is highly complex" (Department of Homeland Security, 2018, p. 5) and that "Software and systems are growing more complex" (The White House, 2023, p. 2). With increased automation and digitalisation in all aspects of our societies, and the growing dependency on technologies that characterise modern socio-political institutions, many predict that this complexity will only increase (Dunn Cavelty & Wenger, 2020). Such complexities are studied primarily as technical problems or policy challenges that need to be fixed or addressed by human actors. This is related to a broader issue in thinking about digital technologies and socio-technical structures, as they are often instrumentalised and viewed in utopian terms when they obey human orders and perform the tasks they are designed for, and in dystopian terms when they do not (Miccoli, 2017; Schandorf & Karatzogianni, 2018).

Contrary to such widespread views, this chapter studies complexities and contingencies, not as problems to be fixed, but rather as intrinsic and ontological properties of information that have generative influences in the construction of cybersecurity. Using the second manifestation of entropy as *randomness*, the chapter shows how entropy as a security analogy can help us understand the security implications of complexities and interdependencies that characterise cybersecurity. The analysis is primarily focussed on codes/software, as a form of syntactic information, and their non-linear operations which the book argues can be theoretically captured through the concept of "emergence." Emergence is a key concept in complexity theory and the study of self-organising systems. It illuminates the inherent unpredictability of complex informational systems and the elements of novelty associated with their operations. As will be shown, the *logic of emergence* challenges the idea of human control in cybersecurity in two ways: by undermining the centrality of human intentionality as a basis for constructing enmity, and by acknowledging the role of the informational non-human in co-producing subjects and objects of cybersecurity.

DOI: 10.4324/9781003454113-5

To illustrate this argument, the chapter is divided into three sections. The first section starts with an exploration of the centrality of complexity in theorising for information and entropy. The second section moves to syntactic information, i.e., codes/software, which malware as the ultimate cyber weapon is a prominent example of. It analyses two main properties of software operations: their entanglement with human and material objects and the elements of contingency and unpredictability in their operation. Lastly, the third section conceptualises cybersecurity as emergent security, in which the complexities of digital information challenge human control of security situations. It introduces this logic of emergence in light of the construction of enmity and the co-production of subjects and objects in cybersecurity discourses and practices.

Information, complexity, and emergence

Like information, complexity is hard to define. It has been often approached in relation to describing, controlling, or building a system or a process in multiple sciences, including computer science, physics, mathematics, among others. In these sciences, complexity has been taken a wide variety of formulations and has been conceptualised in different ways (Burgin & Calude, 2016). Although there is no exact definition for complexity, there is relative agreement on the properties a complex system may possess. These include the existence of various interdependent elements or parts within the system that interact with one another in a non-linear manner. This non-linearity means that the system's behaviour is difficult to predict, and any changes in any of its small parts, however insignificant, may have major implications on the whole system (Merali, 2006). A complex system is also a system that implies the existence of chaos at least in some of its parts. That is why some scholars argue that complexity involves an "interplay between chaos and non-chaos" or that it operates at "the edge of chaos" (Baranger, 2000). Such property leads to what is often referred to as "emerging behaviour."

Emergence is a key concept in complexity theory, which is also linked to cybernetics, computer science, and chaos theory. In one definition, complexity theory is conceptualised as "a science of emergence" (Waldrop, 1993). The key assumption behind emergence is that a complex system will necessarily produce new, unexpected properties and will end up behaving in an unpredictable way (Mason, 2009). As a result of interactions among its diverse parts, the properties of such system will change dynamically in a non-linear process, producing *emergent* rather than *resultant* behaviour. Emergence, non-linearity, and non-equilibrium are characteristics of self-organising and complex adaptive systems, in which outputs cannot be simply predicted based on inputs or by analysing the individual parts of the system. The interactions that take place autonomously in these systems lead to *emergence* (Bousquet & Curtis, 2011).

As can be inferred from the previous chapter, entropy is strongly connected to the concepts of complexity and emergence. Entropy is both a product of and a contributor to complexity. Complex systems are generally non-linear, resulting in random outcomes. From a thermodynamic perspective, more randomness means higher entropy and vice-versa. Further, in some definitions, entropy and its time irreversibility is even described as "an emergent quality of the system" (Standish, 2001, p. 5). Generally speaking, if entropy increases in the system, this results in increasing randomness and unpredictability, hence increasing *emergence* (Johnson et al., 2013). This notion of emergence is capable of countering the reductive assumptions of "ontological individualism" and ideas about humans as the sole agents in the world (Bousquet & Curtis, 2011). It is a statement against an in-control human with a full capacity to understand and predict surrounding environments. In that sense, emergence is seen as *ontological*; it is "real" and does not have to be perceived as such to exist. It is an ontological phenomenon that is fundamental to the operation of self-organising, complex systems (Morçöl, 2013). Accordingly, a shift towards nondeterministic self-organisation theories has been taking place in many fields. These theories study the fluctuations in complex systems and the difficulties of predicting their future state. In such case, the system does not undergo a mechanical transformation that connects causes and effects. The system chooses one among various alternative ways of reactions, without relying on specific structural instructions outside it (Hofkirchner, 2011).

Emergence has a number of characteristics that can be employed in theorising cybersecurity as *emergent and entropic security*. Firstly, emergence is characterised by *novelty*. New features can appear as a result of dynamic changes, which cannot be simply predicted from existing properties of a system. Again, this is connected to non-linearity and unpredictability (Corning, 2002). That is why, dynamical systems theory shifts from trying to understand the exact statistical descriptions in a dynamic system towards analysing non-linear processes instead (Mitchell et al., 1993). Complex, dynamical systems are understood to change over time too and to move towards increasing complexity (Warren et al., 1998).

Secondly, emergence is contextual and relational. Emergent properties in every system are unique to its particular context and to its interactions with multiple agents (Mason, 2009, pp. 32–35). Thirdly, emergence is related to holism: the overall operation of a system is not identical to the behaviour of its self-organising parts. Put differently, the whole cannot be reduced to its parts; i.e., "the whole is bigger than its parts" (Humphreys, 2016, pp. 26–35). Emergent systems are not centralised, and their parts are not necessarily working towards achieving a particular, unified goal. They rather adapt and interact with the dynamic changes in their environments, producing emergent results for the entire system (Corning, 2002). Behaviour that is not completely regular and not completely random is described as "emerging complex behaviour" (Gao et al., 2013).

Information is entirely connected to complexity and the idea of self-organisation. Principally, one way to understand information is sought to be through measuring complexity, and on the other hand, any information obtained about a system aids in decreasing complexity (Burgin & Calude, 2016). In addition, as stated earlier, information is always noisy and prone to error due to problems in transmission, measurements, previous computations, etc. Here, a branch of computational complexity, called information-based complexity, focusses on this kind of partial or noisy information. Due to the absence of sufficient information to solve problems, solutions presented are often "approximate" (Plaskota, 1996).

More practically, the evolution of digital technologies and their growing use have led to increasing connectivity and consequently increasing the speed and volume of information processing, transmission, and storage. This resulted in multiple diverse, networked, and complex systems that demand complex structures with a high level of flexibility and adaptability to manage such complex and dynamic systems. It also resulted in an ever-increasing informational complexity, thus requiring complex computational capacities to handle it (Merali, 2006). Moreover, information systems are inherently "self-organising agents." Self-organisation here can be defined as "the ability of complex systems to spontaneously generate new internal structures and forms of behaviour" (Merali, 2006, p. 220). This self-organisation is not necessarily a design feature; it can spontaneously evolve through the interaction between systems and environments, or between the various parts within the system. Thus, information systems are capable of collecting information and act upon it to pursue a certain set of goals, producing a wide range of future possibilities that cannot be easily predicted (Johnson, 2006).

Therefore, information should be seen as generative of non-formalised "self-determined processes" in complex systems (Hofkirchner, 2011). For that reason, the operation of information technologies and information-processing systems can be only described probabilistically, since it is very difficult to accurately predict their future behaviour (Keyes, 1977). Such complexity challenges human control of information systems, as it is not possible for any one actor to have complete knowledge about the entirety of the system in any given time. As argued by Merali, the state of complex information systems "emerges from the local actions of the individual agents, none of whom have complete knowledge of the entire network" (Merali, 2006, p. 218). This means that complex information systems and networks embody "a spontaneous departure from the past" (Merali, 2006, p. 218), as they do not necessarily follow a predetermined path.

The complexity of codes/software

In popular discourse, terms like codes, software, and algorithms are sometimes used interchangeably as if they all refer to the same thing. For analytical purposes, however, it is useful to examine how each of these terms is technically defined. Generally, there are two types of codes. The first is "source code," which is a textual artefact written by programmers using programming languages, that specify a certain set of instructions that digital devices have to follow to perform their designated functions. These codes are then combined in a form that computers can understand, and thus transformed into "executable code." Software, on the other side, are commercial applications produced by software engineering that transform static codes into processual programs, and in turn act as mediators between codes and real-world execution (Berry, 2011).

As normal users of information technology, we do not interact with codes as an internal, static element of computing systems; rather we interact with software and applications as the external, dynamic element. Codes may be embedded in objects, such as DVDs or computer chips, in infrastructure (like mobile or radio networks), and in processes of information transfer (Kitchin & Dodge, 2005). Codes cannot operate without algorithms; all kinds of codes involve a certain kind of "algorithms and data structures." Algorithms set clear instructions on how a software can operate in order for certain outputs to be produced. They are primarily the ideas that codes aim to execute and they influence every step in software operations, which involves several selections of alternative sets of actions (MacKenzie, 2006).

What is more, codes also embody a high level of complexity. They can be both a solution to a problem when they are produced by software programmers, or the problem itself, when they encounter change in their operation or embody errors that can be exploited in cyber incidents. They are rule-governed but also adapt with the peculiarities of different computing environments. Due to their complexity, codes are sometimes difficult to understand by a single programmer, and in operation, they sometimes seem like as if they are re-writing themselves. They can also be considered as "concealed social orders" (Mackenzie, 2003). On one side, they are designed to control the complexity of the world, but in so doing, they shape our understanding of this complexity and instantiate their own complexities. As Leach argues, sometimes technological artefacts work in harmony with humans, but in other cases, they are stubborn, non-cooperative, and do not meet the expectations of the human creators (Leach, 2020).

The entanglement of codes, humans, and material objects

Power is integral to the logic of bits. The very idea of binary codes is based on an understanding of something as "permitted or not permitted" (Thrift & French, 2002). Even if it is primarily a textual entity, code is more than a "medium of description." Importantly, it is a "medium of execution." This executability and inherent causal power of codes/software is a main property that distinguishes them from other artefacts (Colburn, 1999). Codes/Software are now constitutive of our human existence and are constantly entangled with the functioning of human and material objects. Although "art-like objects" usually have a human recipient, it is not necessarily the case for codes/software. Sometimes the recipients of codes/software are other machines or software, which in turn can generate codes (MacKenzie, 2006).

In cybersecurity, specifically, a malicious software (malware) is targeted towards particular vulnerabilities (exploitable coding errors) in the adversary's system, not humans. Additionally, though they inhabit micro-spaces, codes/software are agents for the "automatic production of space," which is essentially informationalised (Thrift & French, 2002). They play an important role in constructing spatiality in the modern world by controlling, producing, and managing many essential elements of life, be they communications, travel, work, etc. (Kitchin & Dodge, 2005). In cybersecurity, as argued by Blazacq and Dunn Cavelty, malware is capable of co-constructing spatiality by circulating within multiple spaces that cross sovereign boundaries (Balzacq & Dunn Cavelty, 2016). Codes/Software in general create computational ecologies in which humans and non-humans exist. Such ecologies engender "new social ontologies" manifested, for example, in the role social media is currently playing in shaping people's lives. Hence, in many ways, codes/software stand between us and our experience of the world (Berry, 2012).

On one side, codes/software are transforming material objects, increasing their affordance, and stretching their physical limitations. Equipment we use, appliances, medical devices, etc. are all now given the capacity to perform tasks that were not possible before – all thanks to codes/software. Additionally, codes/software can make objects addressable, through bar codes, magnetic stripes, chips, etc. They transform non-digital objects into machine-readable ones, and thus give them an ontological and epistemological unique status by making them traceable across space and time (Kitchin & Dodge, 2011). They create new experiences of spatiality, particularly when coded objects replace non-digital ones entirely. For instance, a supermarket where coded systems fail ceases to function as such because products will not be scanned. Similarly, trains that operate through coded infrastructures are no longer a viable means of transportation if this infrastructure crashes (Kitchin & Dodge, 2005). That is, codes/software alter the capacities of traditional material objects, increase their technicity and affordance, and make them "addressable, aware, and active" (Kitchin & Dodge, 2011).

On the other hand, software and its coded objects, infrastructure, and the processes it enables also influence the way human activities are shaped, regulated, augmented, and facilitated. Humans can now process larger amount of information than ever before, perform complex tasks efficiently, manage systems remotely, engage in new labour practices, etc. (Kitchin & Dodge, 2011). There is always an increasing desire to delegate more to codes/software and to even go further by creating interpersonal relationships between people and their devices. For example, with the development of digital personal assistants, like "Siri" and "Alexa," there is an expectation that technology will figure out our needs without us saying anything (Willson, 2018). This demands a study of the co-evolution of humans and digital objects to explore how just as humans are the creators of the digital, the digital is also influencing their knowledge practices (Adams & Thompson, 2016).

Furthermore, instead of instrumentalising codes/software and studying the extent to which they mirror human intentionality, we should consider elements of contingency in their performance (Rammert, 2012). Computers can repeatedly refuse to follow users' requests by crashing, not opening a file, failing to save it, etc. It is almost always the human who is forced to try to think the same way as a machine to compel it to work in their favour. Humans are capable of adapting to the "stupidities" of a machine much faster than a machine can, because it takes time for updates to be released and for bugs to be fixed. Through such contingencies, humans sometimes are forced to adjust their behaviour to the demands of the machine (Goffey, 2017).

Acknowledging this entanglement of codes/software, humans, and material objects raises multiple ethical questions, particularly in relation to the issue of responsibility (Adams & Thompson, 2016). For this reason, several ethical aspects are now arising with the widespread use of AI and the consequences of the autonomous decisions it makes on equality and social justice. There is a growing acknowledgement of the biases that exist in many algorithms, based on race, gender, or social class; ranging from those responsible for deciding accepted job applicants or granting people loans, to those used in predictive justice and predictive policing. There is also an increasing concern about the use of autonomous weapon systems in warfare, enabled by AI, often called "killer robots." This includes the use of drones and anti-missile defence systems that algorithmically analyse sensor data and make decisions with varying degrees of human intervention. Many movements have already emerged to try and ban the evolution of such systems, in order not to give a non-human system an unsupervised power to make autonomous decisions to kill (Dunn Cavelty et al., 2017; Leese, 2019).

This human vs. non-human responsibility dilemma intensifies due to the secrecy associated with algorithms, or what is referred to as "black-box machine learning." This secrecy allows algorithms to take important decisions in the absence of a thorough understanding from the users' part on how they actually work (Knight, 2017a, 2017b). The resulting questions of responsibility and accountability challenge the liberal-modernist understanding of humans as the sole agency that produces clear causalities based on intentional decision-making (Hoijtink & Leese, 2019). It is thus no surprise that we find an increasing number of literatures talking about the role of codes, not just humans, in producing and reinforcing inequalities (Graham, 2005), perpetuating racism (Sandvig et al., 2016), and other forms of discrimination (Noble, 2018).

The use of AI and machine learning is quite prevalent in cybersecurity practices too (Cristiano et al., 2023). It is estimated that the AI in cybersecurity market reached USD 22.49 billion in 2023 and is forecasted to grow to USD 114.30 Billion by 2031 (SkyQuest, 2024). DARPA has introduced a research programme in 2019 entitled "Guaranteeing AI Robustness against Deception" in order to advance deception-resistant machine learning systems that could defend against AI attacks (Taddeo et al., 2019). This is not new; as explained by Stevens, using algorithms as a method of intelligence gathering for cybersecurity goes back to the 1990s (Stevens, 2020). Nowadays, malware, vulnerability, and intrusion detection processes are moving towards more automation. AI adds an operational advantage to cybersecurity strategies since it is capable of overcoming the limited cognitive abilities of humans to handle huge amounts of data. This creates new epistemic assemblages in cyber defence that combine humans, machines, and algorithms. In these assemblages, threat intelligence becomes a process that is not only performed by humans, but is rather one in which the non-human AI produce information that shape security decision-making (Stevens, 2020).

Because AI systems are inherently dynamic, understanding their operation and explaining their outcomes is not an easy task. That is why, when AI systems are attacked, detection becomes difficult, because reverse-engineering their operation to understand their behaviour is quite challenging. This makes it difficult to know whether the outcome of such behaviour is a result of an attack or not (Taddeo et al., 2019). Not only is AI growing in use in defence practices, but also in offence. With increasing AI capabilities that do not require a lot of human labour or intelligence gathering, the costs of cyber operations are being lowered. Experts predict that new type of cyber incidents are likely to appear in the future given that AI is capable of transcending what humans may consider impractical, such as labour-intensive spear phishing operation (Brundage et al., 2018).

Contingency, non-linearity, and unpredictability

Codes as textual objects are distinguished from normal language by their "executability." They do not depend on externalities and do not require the same level of mediation as language. Once embedded in a digital machine, codes start operating automatically, telling that machine what to do or not to do, sometimes without requiring any human intervention, especially in the case of ordinary users of technology (Frabetti, 2015). In many tasks, starting from simply logging into the internet, codes/software react to inputs and outputs automatically, often with no direct human intervention (Kitchin & Dodge, 2011). Machines are now automatically exchanging data, using electronic sensors, updating themselves, producing predictions and warnings, controlling traffic lights, authorising payment cards, opening and closing doors, etc. (Berry, 2011). Starting from the year 2008, the number of "things" on the internet even exceeded that of humans on Earth; a trend that amplified with the advent of the internet of things (Hansen, 2017).

Computer-mediated information processing embodies some form of intelligence and is capable of interfering in many human tasks such as memory and cognition. They are also more malleable, flexible, adaptable, and interactive to the outside world than other technologies and "material artefacts" (Kallinikos, 2010). Now, algorithms and AI are taking decisions for humans in many fields, with little to no communication on how and why they chose a particular course of action. Humans are sometimes faced with a condition in which they have to either trust the machine and follow its choices or not use it at all.

Instead of humans being in absolute control of the digital, they are now constantly tracked by machines and sensors that are collecting data about them, sometimes without their knowledge or permission (Norman, 2009). The rise of web beacons is one obvious example in this regard. These are automated agents for data collection, composed of algorithms that are presented in the form of small one-pixel graphical images embedded on websites and browsers – so small that users cannot see. These beacons are capable of constantly collecting data about users, influencing their behaviour, and tampering with their actions online. Although many mechanisms are being developed to allow users to know who is tracking their data, these web beacons are growing more complex and difficult to understand, even on the part of programmers (Berry, 2014).

Further, there are multiple forms of unpredictability associated with codes/software starting with the way they are produced. Codes/software are not developed by a single person but are usually engineered within big projects in which many programmers with varying levels of skills and knowledge participate. This process results in a very complex piece of software that no one single programmer can claim they fully understand (Kitchin & Dodge, 2011). The more a software remains "alive," the larger the number of programmers involved in its development and maintenance, and the more difficult it is for a single programmer to fully understand its complex operation. Further, in most cases, software is engineered through a process of trial and error. They are left to run and have a life of their own, while being tested and improved in the process. That is why, software is mainly *engineered* rather than *designed*, since it does not always follow what programmers dictate. In such case, programmers almost have an "ignorant expertise" in dealing with codes/software they helped producing (Thrift & French, 2002).

Hence, the operation of codes/software is never linear; they usually incur self-modification and deviation in execution. This deviation is also common due to bugs (errors) that are likely to occur regardless of how efficient code-writing is (Chun, 2011). This is what one author described as "code drift" to explain the many unplanned consequences, fluctuations, and transformations that occur in the operation of codes/software. It can be argued thus that albeit the illusion of control, information systems are evolving as "code drifters" (Kroker, 2014). Added to this, programming is done by standardised, formalised software-enabled languages that facilitate the process of writing code. This involves a lot of abstractions that hide details that may seem unnecessary for the programming process. Although these abstractions make the job of programmers a lot easier and enable people to code even if they do not possess sufficient technical capacities, they also reduce their knowledge of and power over the codes they write. As argued by one study, automatic programming "is both an acquisition of greater control and freedom, and a fundamental loss of them" (Chun, 2011, pp. 45–46).

This tendency is magnified when it comes to normal users who are neither given the access to such codes, nor do they have the required knowledge to understand them. Ordinary users normally have no comprehension of internal codes and algorithmic processes beyond the graphical interfaces they interact with. Such interfaces give the human user an illusion of control and an imaginary of a "sovereign executive," when in fact they are perpetuating users' ignorance (Chun, 2011). The ignorance on the part of users is influenced by two main principles in the design of digital objects: encapsulation and exception handling. Encapsulation refers to the way through which programmers hide code implementation from users so that they prevent code tampering that may cause errors. Exceptional handling refers to the way machines are programmed to deal with surprises and problems without necessarily consulting the user or informing them of the problem. Although both are meant to facilitate users' experience in dealing with digital objects, but they also significantly decrease their knowledge of the system (Goffey, 2017).

It can be argued, therefore, that codes/software are in constant state of emergence. They are designed to interact with their environment, with little or no direct intervention from humans. For instance, the algorithms of page ranking on Google, or post rating on Facebook, interact with different users differently, based on their own individual interests. They use non-fixed codes designed to evolve by themselves through a process of independent learning while interacting with the user. In some cases, algorithms are built to be random and unpredictable. One example is the algorithm used in autocomplete in Google search that produces different results when the same letters are typed in different contexts (Kitchin, 2018). As opposed to rule-based algorithms in which humans specify clear instructions to be followed for producing a certain output, machine-learning algorithms give the machine a certain set of data, accompanied by some feedback, and then leave it to operate independently to determine the best way to reach an output. Although machine-learning algorithms have the advantage of solving many problems that humans cannot possibly write instructions for, they also make it hard for programmers to understand the steps they took to reach a solution (Fry, 2018).

Beyond that, codes/software also operate under different temporalities that are much quicker than that of a human. We usually only notice codes/software exist when they slow down, stop working, or when there is a glitch (error) in the system. This is what Berry described as "the glitch ontology" (Berry, 2014). In interacting with software, the ordinary user cannot possibly know what their actions will result in or if the result is what they intended. Also, the computational device is in itself a mediator between objects and their representation. Anything can be changed, altered, and manipulated in many ways before being represented to the user. For example, using a telescope to see a microbe, there is first a conversion of the analogue microbe into a digital image saved in the memory of the computer as numbers or bits. In processing this image, the computer's software can manipulate it in different ways, like changing its size, colour combinations, etc. Then the processed representation is introduced to the user after transforming the binary digits to an image that they can see. The ordinary user's control on this process of transformation is very limited (Berry, 2011).

Finally, the unpredictability and contingencies of codes/software are even more prominent when they are used maliciously. Several unintended political and technical consequences that transcend the control of the initiator may result from self-replicating malwares. They can spread to un-targeted systems, they can be discovered due to an error in coding, and they can cause an over-reaction from governments or media that was not initially intended. Assuming that objects also enact spaces, it can be argued malware is co-constitutive of the "space" in cyberspace, meaning that cyber incidents should be analysed within "the spaces they build themselves" when spreading between devices in unplanned ways by their initiators (Balzacq & Dunn Cavelty, 2016). This argument has far-reaching implications on security and the assumptions of human control imbedded in its logics, as will be explained next.

The logic of emergence and human control in entropic security

As explained in more detail in Chapter 3, the logic of emergency in security studies embodies an implicit assumption on human control of security environments. It is humans who construct security threats as existential and they are the ones who implement exceptional, emergency measures in response. As argued by critical security scholars, this logic is intrinsically linked to enmity and direct causal relations to perceived threats. Here, constructing enmity is assumed to be a human choice with the purpose of invoking security. Arguably, this assumption is problematic because it does not account for the contingencies and unpredictability intrinsic to information operations and therefore to cybersecurity. Assuming that emergency measures are integral to security construction means that their introduction and implementation are always a possibility. It implies that it is up to humans' desires and intentionality to propose and implement such measures. This becomes problematic if the above-mentioned contingencies of codes/software are considered in studying cybersecurity, particularly in the case of malwares.

Malwares – commonly also referred to as "cyber weapons" – are special kinds of codes/software. Cyber incidents caused by malwares are major challenges to ideas of control upon which cybernetics and computing technologies were based. It is the dystopia of the promises of "cybernated" economies, cyborgs, and cyberspace as a new parallel frontier to reality. Now, control over machines can be taken from humans, systems can be attacked and controlled distantly, and several damages can result in the form of data loss, abuse, denial of services, or even machine damaging (Rid, 2016). Malwares represent the uncontrollable forces that challenge the idea of user's control over a system as a function of its security. The relative unpredictability of malwares can challenge human control, even if operating through rationally pre-defined codes and algorithms (Parikka, 2007).

The most peculiar property of viruses and worms is not their maliciousness, because they are not malicious per se, but rather their ability to copy themselves automatically, described as "self-reproducing automata" (Parikka, 2017). While viruses require human action to activate them, like clicking a link or opening a file, worms even have the capacity to self-propagate. Malwares are

also capable of performing multiple self-preservation techniques that complicate security measures. For example, they can do what is known as stealthing,
through which they hide their presence and make it difficult to detect them.
This can be done by slowing down their operation or presenting a fake clean
image of an infected file to an anti-virus program. Polymorphism is another
self-preservation technique, by which malwares change their base code dynamically every time they run, whilst having the same functionality. A step further
to polymorphism is metamorphism, which refers to malwares changing their
functionality as they propagate across different systems (Skoudis & Zeltser,
2004). In short, malwares are inherently active; they are constantly doing
something or spreading somewhere, "almost like living" (Parikka, 2017).

Complex, dynamic, and decentralised information systems with emergent
behaviour produce complex, dynamic, and decentralised security with emergent properties. The elements of unpredictability in the operation of codes/
software as described earlier generate a *logic of emergence* in cybersecurity. The
book proposes the logic of emergence as an alternative way of theorising
cybersecurity, and as the second logic in the trilogy that constitutes entropic
security that goes beyond emergency measures. To be clear, this does not
entirely invalidate human control in cybersecurity. Rather, it suggests that the
construction of security in cybersecurity is not always subject to the sole intentionality of humans. The elements of novelty, unpredictability, contextuality,
and decentralisation associated with emergence can be found in the production
of enmity and the subjects and objects of cybersecurity, as will be explained in
the next two subsections.

The subjects and objects of cybersecurity

Cybersecurity is distinguished by its multi-stakeholder nature. It is co-
constituted by every single user of digital technologies, from individual citizens
to corporations and governments. However, the identification of the actors of
interest in a certain incident and those entitled with taking the necessary measures to counter an ongoing cyber incident or attack is not always pre-defined
and can have an emergent nature. Similarly, choosing security *objects* in a
single incident may not also be entirely controlled by the attacker. The subjects
and objects of cybersecurity, together with the resulting consequences of a
cyber incident, are co-produced by malware *and* human actors.

Firstly, if all software contains bugs (coding errors), a malware is distinct
given its ability to self-replicate, which intensifies its potential buggy nature.
Bugs are more likely to appear in malware because they do not go through the
same testing processes of normal software. Further, since malware does not
operate in controlled environments, it becomes difficult to overrule bugs they
may contain during propagation, and therefore increasing chances of unintended consequences. Once a malware is deployed, it becomes very difficult for
the attacker to maintain control over its propagation or to accurately predict
its behaviour. It can always affect unintended systems resulting in varying

degrees of damage. Not knowing strictly which systems the malware will propagate to beforehand, in most cases, limits the attacker's ability to test its compatibility with such systems (Cobb & Lee, 2014).

Secondly, even if propagation is meant to be limited, in practice, that might not be possible, particularly because attacks can hardly be stopped once started. To reach its target faster, the attack needs to spread widely and to propagate fast among non-target systems. A specific algorithm is usually used for target selection, either by simply choosing random IP addresses to infect,[1] or target neighbouring devices on the same local network as the victim. Once on the target's system, those algorithms can also choose other targets from email address books, DNS server, among other ways (Panagiotis, 2006). This relative unpredictability of malware is one reason why some scholars criticise the use of cyber-attacks by states as a purportedly more ethical choice than military attacks. They argue that unintended and uncontrollable potential implications of cyber-attacks on civilian targets make the argument about their ethical use obsolete (Rowe, 2017). For the same reasons, some argue that collateral damages in cyber-attacks are even much higher than military attacks (Hirsch, 2018).

There are numerous examples that demonstrate the inaccuracy of cyber targeting, leading to unintended consequences. The NotPetya ransomware of 2017 is thought to have been targeted at companies in Ukraine. However, the target verification mechanisms of the ransomware did not work properly, and it ended up infecting a large number of targets far away from Ukraine and in several parts of Western Europe. Another example is an attack that exploited a vulnerability in a software called CCleaner. Due to an error in coding, the attack ended up infecting targets in Slovakia instead of its initial target: South Korea. But perhaps the most notable example in this regard is Stuxnet worm that, as widely believed now, was designed by the United States and the Israeli governments to target the Iranian nuclear centrifuges in 2010. The worm was imbedded on the targeted system initially using a USB stick, before it started propagating. Stuxnet spread to multiple other unintended targets outside Iran, including Germany, China, and even the USA itself. This happened despite the high level of sophistication of this worm, which many believe was designed over many years. It is thought to have included some methods of limitation that developers used to curb its wide proliferation. But these anti-propagation measures and complex design did not stop it from producing unintended consequences. Though it had a specific target, it transmitted to more than 100,000 computers in various locations in its original propagation (Hirsch, 2018). The spokesman of Chevron, an American multinational energy corporation that was hit by Stuxnet reportedly said upon discovering the malware in the company's systems: "I don't think the U.S. government even realized how far it had spread…I think the downside of what they did is going to be far worse than what they actually accomplished" (King, 2012).

Hence, accidental and unpredictable damages do happen, whether for reasons related to contingencies of codes/software as outlined earlier, or because of the use of legacy codes that are old enough to not be supported by modern software/hardware, even in CNIs (Backman, 2023). It could be argued that although the threat actors behind cyber incidents can choose which software/hardware vulnerability to exploit, and in turn which private actor would need to issue patches to stop an attack, many elements of the cybersecurity environment become *emergent*. As argued by Dwyer, cyber-attacks should not be studied as linear relationships between hacker intents and resultant impacts through malware as tools (Dwyer, 2021). Even with the existence of targeting mechanisms, the contingencies of malware per se co-determines which systems gets infected at the end-users' side during its propagation. By propagating across machines, malware co-creates a network of cybersecurity actors who are then required to take steps to stop the attack, such as updating their systems to apply the necessary patches. In doing so, malware contribute to emergent, contextual actancy in every single incident.

For instance, in 2017, the WannaCry ransomware exploited a vulnerability in Microsoft operating system that allowed for remote execution of a code that encrypts files in the infected systems. The choice of the infected targets depended entirely on the agency of the malware during self-propagation, by scanning unpatched systems and deploying itself. It reportedly infected more than 230,000 systems in 150 countries, among which was the National Health Service (NHS) in the UK (Cooper, 2018). By infecting its systems, the malware put the NHS under the spotlight as a major cybersecurity actor. Much of the blame was directed towards the entity for not updating its systems to apply the patch issued by Microsoft before the attack (BBC, 2017). This has also raised scholarly and policy interest in the importance of cybersecurity for health care. As argued by Ciaran Martin, the former head of the UK's National Cyber Security Centre (NCSC): "WannaCry and NotPetya were deliberate attacks, but their impact on the UK and allied countries was accidental. So the two biggest incidents that we faced early on [at the NCSC] were both basically accidents" (Kelly, 2020).

This does not only apply to big entities, but also to individual users who become influential actors when a particular attack takes place and infects their machines, as well as to other objects. An important example here is the Mirai malware which was designed to target IoT devices and make them part of a botnet (a network of other compromised devices) to launch a distributed denial of service (DDoS) attack in 2016. Millions of users were not able to connect to various websites as a result of this attack.[2] The Mirari botnet, as argued by Liebetrau and Christensen, constituted a "dance of agency," in which malware constantly moved in ways that were not entirely predictable, transforming mundane IoT entities in 164 countries into bots with damaging effects (Liebetrau & Christensen, 2021).

These contingencies and elements of unpredictability of syntactic information, thus, undermine the idea of an in-control human actor who manages cybersecurity environments. It is a demonstration of the influence of codes/software in co-producing cybersecurity, which in turn becomes *emergent security*. Furthermore, these properties syntactic information create a liability and responsibility dilemma in cybersecurity that resembles the prominent risk theorist Ulrich Beck's argument on the second modernity and its "highly differentiated division of labour" that results in a "general complicity" and lack of responsibility in the production of risk. As Beck said, "Everyone is cause and effect, and thus non-cause" (Beck, 1992, p. 33). But this dilemma in cybersecurity is not specifically just a result of modernity. Rather, as argued in this book thus far, it is primarily co-produced by the ontology of information.

One implication of conceptualising cybersecurity as *emergent security* is problematising "active cyber defence" or "defend forward" cybersecurity strategies by governments. These operations may include non-disruptive practices like hacking adversaries' or allies' information systems and maintaining a presence in such systems for intelligence gathering, or disruptive operations like "hacking back" to recover stolen data, for instance (Healey, 2019; Pattison, 2020). Acknowledging the complexities and contingencies of codes/software and the elements of unpredictability and uncertainty in their operation challenges the idea of human control implicit in such understandings of "cyber defence." As shown above, a malware used in targeting a certain system can spread to untargeted ones, even within the geographical location of the initiator, and therefore result in several unintended consequences. That is to say, although some states may conduct cyber intrusions with defensive motives in the background, acknowledging the agency of codes/software illuminates the risks of condoning such operations by labelling them "defensive."

Enmity and the attribution dilemma

Establishing an enemy in cybersecurity is a complicated process that does not just reflect the actors' decisions, but also the complexities and contingencies of codes/software. Enmity in cybersecurity, just as complexity is described by one of its early writers, "…arises out of the combined agencies, but in a form which does not display the agents in action" (Lewes, 1875, p. 368). Elements of novelty, non-linearity, contextuality, and decentralisation are manifested in constructing enmity in multiple ways. Firstly, malware conditions the centrality of human intents in constructing cyber threats. This is because hostile intents and aggressors' capabilities are not the only deciding factors for the occurrence and success of a cyber-attack. For a cyber-attack to take place, a vulnerability has to be identified in the targeted system first; and security vulnerabilities are essentially contextual: they vary across different systems. Besides, the implications of cyber incidents are mainly linked to the level of the target's dependency on information systems. The less cyber dependent the target is, the less effective an attack against it would be, making the impact of such an attack relational too. That is why, it is argued that in cybersecurity, "offensive capacity correlates with defensive vulnerability" (Schutte, 2012, p. 8). Put differently, human intentionality is not enough to launch a cyber-attack.

Secondly, cybersecurity is characterised by a high level of asymmetries between actors and their capabilities that often render any attribution-specific defence strategy insufficient. As argued by one study, "Whereas defenders in the physical domain can reasonably assume that pretty criminals do not have nuclear weapons and that foreign military powers will not rob the local McDonald's, this same categorical logic does not hold true in cyberspace"(Rivera & Hare, 2014, p. 104). That is, attack sophistication is not necessarily evidence for state-sponsorship. Added to that, determining the cyber capabilities of a certain actor is often more a matter of speculation than knowledge. Unlike military arms, the non-physicality of cyber offensive tools makes them almost unobservable, unquantifiable, and in most cases, unrecognisable before an attack actually takes place (Schutte, 2012). This, in turn, puts more emphasis on codes/software than human aggressors in immediate cyber defence. It is coding vulnerabilities and exploits used to target them that lie at the core of such defence, even when enmity is more discursively prevalent.[3]

Thirdly, the complexities and contingencies of codes/software challenge attack attribution even further, making it primarily a process driven by profound uncertainties. For instance, malware may take control of a user's computer without their knowledge, creating a network of devices that work together to orchestrate an attack in a way that crosses geographical boundaries. The malware moves between devices across borders, scanning for the targeted vulnerability without consulting the attacker on the devices it affects. This makes it difficult to know if a certain device is acting as a bot or not and to determine who is controlling it, particularly given the irrelevance of geographical proximity as an element of attribution (Singer & Friedman, 2014). This also means that systems can be hijacked by a third party to implant attacks.

Accordingly, attribution is not necessarily part of an immediate response to counter cyber-attacks. Although publicly published reports on attack attribution by the private sector exceed those of governments (Rid & Buchanan, 2015), it is governments and some think tanks that focus more on *threat* attribution.[4] More specifically, intelligence communities are generally more concerned with attributing cyber threats to a particular enemy than private operators and defenders of information systems. This can be seen, for example, in the US government's emphasis on nation-states as a threat source, namely Russia, China, Iran, and North Korea. However, on the practice-level and in immediate responses against a hostile cyber operation, the logic of enmity is not so central. The emergent properties of codes/software condition enmity, particularly in everyday cybersecurity that does not necessarily get publicised (Fouad, 2021; Slupska, 2019). Thus, the enemy is not just a human attacker or a particular actor; *the enemy also becomes the vulnerability and the malware: codes/software.* This can be seen in the technologies and threat mitigation policies developed for cyber defence which are primarily focussed on the tools that adversaries use in hostile cyber operations, rather than determining who this adversary actually is. As argued by a representative of a network security company: "intelligence and law enforcement entities often prioritize attack attribution, while almost no emphasis is placed on attribution by those defending systems" (Reviewing the Federal Cybersecurity Mission, 2009, p. 27).

Acknowledging that the immediate adversary in cybersecurity is the vulnerability and the malware, i.e., codes/software, calls for prioritising the production of secure-by-design codes over instrumentalising such codes in hacking into other countries' systems as part of defensive strategies. In addition to the above-mentioned active cyber defence strategies, states are increasingly involved in black markets of vulnerabilities and zero-day exploits to build their cyber arsenals. Such practices increase the market price of those vulnerabilities and exploits which may end up in the wrong hands and undermine the long-term security of individual users and their overall trust in technology (Herzog & Schmid, 2016). On the other hand, software manufacturers rush production processes to get their products into the market fast enough to compete for profits with an intention to "fix vulnerabilities later." They also tend to prioritise functionality over security in software production, leading to a general culture of acceptance of software insecurity (Chong, 2016). Although cybersecurity is entropic and there will always be *another* bug in every software, shifting the conceptualisation of the adversary in cybersecurity from the attacker to vulnerabilities and malware (i.e., codes/software) challenges such practices by state and private actors and emphasises the dangers of weaponising codes in cyber operations that implicitly assume a high level of human controllability.

Conclusion

This chapter discussed the complexities and contingencies of codes/software as one peculiar property of information and analysed their implications on cybersecurity construction. Instead of instrumentalising information technologies or analysing them as a mere capability that influence power relations among actors in international politics; the chapter focussed on the properties of information in and of itself. It interrogated the ontology of codes/software, their intrinsic complexities, and how they are entangled with humans and non-human objects in cybersecurity and beyond. This property of digital information has significant ramifications for the logics of enmity and emergency measures as widely applied in cybersecurity literature.

Contrary to underlying assumptions of human control that are embedded in the logic of emergency in security studies, this chapter proposed *emergence* as an alternative logic of security. Theorising cybersecurity as entropic security that is co-governed by the logic of emergence acknowledges the self-organising, dynamic, and complex nature of digital information systems and their role in co-producing discourses, policies, and logics of cybersecurity. The chapter argued that the non-linearity and unpredictability produced by such properties can challenge conventional ideas about enmity and the construction of subjects and objects of security. This necessarily produces a kind of security that is in itself non-linear and emergent; i.e., *entropic*.

Codes/software assume a central position in cybersecurity, given their role as the ultimate cyber "weapon" in the form of malwares, and as the embodiment of vulnerabilities that these malwares target. Codes/software can operate in ways that are unpredictable by their creators and generate contingencies in their interaction with humans and machines. Codes/software also operate in digital habitus that humans find themselves having to comply with in many cases, and they are capable of granting technicity material objects in ways that may stretch their physical properties. They co-construct spatiality in the modern world and create computational ecologies in which other agents exist.

The complexity and mailability of codes/software has the capacity to co-generate a different logic of enmity in cybersecurity. Self-replicating malwares that propagate beyond the aggressor's control, the use of bots in attacks without users' knowledge, the packets that change multiple times before reaching the target, and many other factors make it difficult to establish attribution in cybersecurity. Also, in direct cyber defence in the face of an incident, it is usually the code that is the threat, not the human enemy. This complicates the logic of enmity and undermines its centrality in practices of cybersecurity, even if it is more prevalent on the discursive level, particularly in relation to the military and intelligence communities. Furthermore, even when human aggressors make the initial choice of targets, accidental damages do occur, thus determining which other targets are affected by attacks and which actors are important in the line of defence. Accordingly, the logic of emergence becomes a more accurate description of the way threats and security practices unfold in cybersecurity – given its intrinsic relation to the properties of information – than the logics of exceptionality and emergency in security studies.

Notes

1 IP stands for internet protocol. An IP address is a unique address that identifies a device on the internet or a local network.
2 DDoS attacks flood computer servers with requests to stop them from providing services to their intended users.
3 This argument particularly refers to passive defence, or defence that happens after an incident takes place, in contrast to active defence or what is often called "defend forward," which takes pre-emptive actions by intruding in the adversaries' systems.
4 We can differentiate between two types of attribution: attack attribution and threat attribution. The first is concerned with attacks that have already taken place, while the second is related to ones that have not and thus seeks to establish links between the future threat/hazard and a particular source.

References

Adams, C., & Thompson, T.L. (2016). *Researching a Posthuman World: Interviews with Digital Objects*. Springer.
Backman, S. (2023). Normal Cyber Accidents. *Journal of Cyber Policy*, *8*(1), 114–130.
Balzacq, T., & Dunn Cavelty, M. (2016). A Theory of Actor-Network for Cyber-Security. *Reviewing the Federal*, *1*(02), 176–198.

Baranger, M. (2000). Chaos, Complexity, and Entropy. *Ew England Complex Systems Institute, Cambridge.* https://www.cs.auckland.ac.nz/~cristian/UMCreadings/cce.pdf

Beck, U. (1992). *Risk Society: Towards a New Modernity.* SAGE Publications Ltd.

Berry, D. (2011). *The Philosophy of Software: Code and Mediation in the Digital Age.* Springer.

Berry, D.M. (2012). The Social Epistemologies of Software. *Social Epistemology, 26*(3–4), 379–398.

Berry, D.M. (2014). *Critical Theory and the Digital.* A&C Black.

Bousquet, A., & Curtis, S. (2011). Beyond Models and Metaphors: Complexity Theory, Systems Thinking and International Relations. *Cambridge Review of International Affairs, 24*(1), 43–62.

Brundage, M., Avin, S., Clark, J., Toner, H., Eckersley, P., Garfinkel, B., Dafoe, A., Scharre, P., Zeitzoff, T., & Filar, B. (2018). *The Malicious Use of Artificial Intelligence: Forecasting, Prevention, and Mitigation.*

Burgin, M., & Calude, C.S. (Eds.). (2016). *Information And Complexity.* World Scientific.

Chong, J. (2016). Bad Code: Exploring Liability in Software Development. In R. Harrison, T. Herr, & R.J. Danzig (Eds.), *Cyber Insecurity: Navigating the Perils of the Next Information Age* (pp. 69–86). Rowman & Littlefield Publishers.

Chun, W.H.K. (2011). *Programmed Visions: Software and Memory.* MIT Press.

Cobb, S., & Lee, A. (2014). *Malware is Called Malicious for a Reason: The Risks of Weaponizing Code.* 71–84. https://ccdcoe.org/uploads/2018/10/CyCon_2014.pdf#page=84

Colburn, T.R. (1999). Software, Abstraction, and Ontology. *The Monist, 82*(1), 3–19.

Cooper, C. (2018, May 16). *WannaCry: Lessons Learned 1 Year Later.* https://www.symantec.com/blogs/feature-stories/wannacry-lessons-learned-1-year-later

Corning, P.A. (2002). The Re-Emergence of "Emergence": A Venerable Concept in Search of a Theory. *Complexity, 7*(6), 18–30.

Cristiano, F., Broeders, D., Delerue, F., Douzet, F., & Géry, A.. (Eds.).(2023). *Artificial Intelligence and International Conflict in Cyberspace.* Taylor & Francis.

Department of Homeland Security. (2018). *Cybersecurity Strategy.* https://www.dhs.gov/sites/default/files/publications/DHS-Cybersecurity-Strategy_1.pdf

Dunn Cavelty, M., Balzacq, T., & Fischer, S.-C. (2017). Killer Robots' and Preventive Arms Control. In *Routledge Handbook of Security Studies* (pp. 457–468). Routledge.

Dunn Cavelty, M., & Wenger, A. (2020). Cyber Security Meets Security Politics: Complex Technology, Fragmented Politics, and Networked Science. *Contemporary Security Policy, 41*(1), 5–32.

Dwyer, A.C. (2021). Cybersecurity's Grammars: A More-Than-Human Geopolitics of Computation. *Area.* https://doi.org/10.1111/area.12728

Fouad, N.S. (2021). Securing Higher Education Against Cyberthreats: From an Institutional Risk to a National Policy Challenge. *Journal of Cyber Policy, 6*(2), 137–154.

Frabetti, F. (2015). *Software Theory: A Cultural and Philosophical Study.* Rowman & Littlefield International.

Fry, H. (2018). *Hello World: How to be Human in the Age of the Machine.* Penguin Random House UK.

Gao, J., Liu, F., Zhang, J., Hu, J., & Cao, Y. (2013). Information Entropy as a Basic Building Block of Complexity Theory. *Entropy, 15*(9).

Goffey, A. (2017). The Obscure Objects of Object Orientation. In M. Fuller (Ed.), *How To Be a Geek: Essays on the Culture of Software* (pp. 15–36). John Wiley & Sons.

Graham, S.D.N. (2005). Software-Sorted Geographies. *Progress in Human Geography, 29*(5), 562–580.

Hansen, H.K. (2017). What Do Big Data Do in Global Governance? *Global Governance: A Review of Multilateralism and International Organizations, 23*(1), 31.

Healey, J. (2019). The implications of persistent (and permanent) engagement in cyberspace. *Journal of Cybersecurity*, 5(1), tyz008.

Herzog, M., & Schmid, J. (2016). Who Pays for Zero-Days? Balancing Long-Term Stability in Cyber Space Against Short-Term National Security Benefits. In K. Friis & J. Ringsmose (Eds.), *Conflict in Cyber Space: Theoretical, Strategic and Legal Pespectives* (pp. 97–114). Routledge.

Hirsch, C. (2018). Collateral Damage Outcomes are Prominent in Cyber Warfare, Despite Targeting. In L. Leenen (Ed.), *ICCWS 2018 13th International Conference on Cyber Warfare and Security* (pp. 281–286). ACPIL.

Hofkirchner, W. (2011). Does Computing Embrace Self-Organization? In G.D. Crnkovic & M. Burgin (Eds.), *Information and Computation: Essays on Scientific and Philosophical Understanding of Foundations of Information and Computation* (pp. 185–202). World Scientific.

Hoijtink, M., & Leese, M. (2019). How (not) to Talk About Technology: International Relations and the Question of Agency. In M. Hoijtink & M. Leese (Eds.), *Technology and Agency in International Relations* (pp. 1–24). Routledge.

Humphreys, P. (2016). *Emergence*. Oxford University Press.

Johnson, J. (2006). Can Complexity Help Us Better Understand Risk? *Risk Management*, 8(4), 227–267.

Johnson, J.J., Tolk, A., & Sousa-Poza, A. (2013). A Theory of Emergence and Entropy in Systems of Systems. *Procedia Computer Science*, 20, 283–289.

Kallinikos, J. (2010). *Governing Through Technology: Information Artefacts and Social Practice*. Springer.

Kelly, R. (2020, October 7). Emerging Cyber Threats and Their Unintended Consequences. *Digit*. https://digit.fyi/ciaran-martin-ncsc-unintended-consequences-cybersecurity/

Keyes, R.W. (1977). Physical Uncertainty and Information. *IEEE Transactions on Computers*, C–26(10), 1017–1025. https://doi.org/10.1109/TC.1977.1674737

King, R. (2012, November 8). Stuxnet Infected Chevron's IT Network. *The Wall Street Journal*. https://blogs.wsj.com/cio/2012/11/08/stuxnet-infected-chevrons-it-network/

Kitchin, R. (2018). Thinking Critically About and Researching Algorithms. In D. Beer (Ed.), *The Social Power of Algorithms* (pp. 14–29). Routledge.

Kitchin, R., & Dodge, M. (2005). Code and the Transduction of Space. *Annals of the Association of American Geographers*, 95(1), 162–180.

Kitchin, R., & Dodge, M. (2011). *Code/space: Software and Everyday Life*. MIT Press.

Knight, W. (2017a, July 12). Biased Algorithms Are Everywhere, and No One Seems to Care. *MIT Technology Review*. https://www.technologyreview.com/s/608248/biased-algorithms-are-everywhere-and-no-one-seems-to-care/

Knight, W. (2017b, October 3). Google's Ai Chief Says Forget Elon Musk's Killer Robots, and Worry About Bias in Ai Systems Instead. *MIT Technology Review*. https://www.technologyreview.com/s/608986/forget-killer-robotsbias-is-the-real-ai-danger/

Kroker, A. (2014). *Exits to the Posthuman Future*. John Wiley & Sons.

Leach, T. (2020). *Machine Sensation: Anthropomorphism and 'Natural' Interaction with Nonhumans*. Open Humanities Press.

Leese, M. (2019). Configuring Warfare: Automation, Control, Agency. In M. Hoijtink & M. Leese (Eds.), *Technology and Agency in International Relations* (pp. 42–65). Routledge.

Lewes, G.H. (1875). *The Principles of Certitude. from the Known to the Unknown. Matter and Force. Force and Cause. the Absolute in the Correlations of Feeling and Motion. Appendix: Imaginary Geometry and the Truth of Axioms. Lagrange and Hegel: The Speculative Method. Action at a Distance*. Trübner & Company.

Liebetrau, T., & Christensen, K.K. (2021). The Ontological Politics of Cyber Security: Emerging Agencies, Actors, Sites, and Spaces. *European Journal of International Security*, 6(1), 25–43.

Mackenzie, A. (2003). The Problem of Computer Code: Leviathan or Common Power. *Institute for Cultural Research, Lancaster University*. https://s3.amazonaws.com/academia.edu.documents/30698019/code-leviathan.pdf?AWSAccessKeyId=AKIAIWOWYYGZ2Y53UL3A&Expires=1555602968&Signature=4qWZML52LAXvW679MqYHxUGlJgM%3D&response-content-disposition=inline%3B%20filename%3DThe_problem_of_computer_code_Leviathan_o.pdf

MacKenzie, A. (2006). *Cutting Code: Software and Sociality*. Peter Lang Inc., International Academic Publishers.

Mason, M. (2009). *Complexity Theory and the Philosophy of Education*. John Wiley & Sons.

Merali, Y. (2006). Complexity and Information Systems: The Emergent Domain. *Journal of Information Technology*, 21(4), 216–228.

Miccoli, A. (2017, September 27). Posthuman Suffering. *Critical Posthumanism Network*. http://criticalposthumanism.net/posthuman-suffering/

Mitchell, M., Crutchfield, J.P., & Hraber, P.T. (1993). *Dynamics, Computation, and the 'Edge of Chaos': A Re-Examination* (arXiv:adap-org/9306003). arXiv. https://doi.org/10.48550/arXiv.adap-org/9306003

Morçöl, G. (2013). *A Complexity Theory for Public Policy*. Routledge.

BBC. (2017, October 27). NHS Trusts 'at Fault' Over Cyber-Attack. https://www.bbc.com/news/technology-41753022

Noble, S.U. (2018). *Algorithms of Oppression: How Search Engines Reinforce Racism*. NYU Press.

Norman, D. (2009). *The Design of Future Things*. Hachette UK.

Panagiotis, K. (2006). *Digital Crime and Forensic Science in Cyberspace*. Idea Group Inc (IGI).

Parikka, J. (2007). *Digital Contagions: A Media Archaeology of Computer Viruses*. Peter Lang.

Parikka, J. (2017). *Digital Contagions: A Media Archaeology of Computer Viruses*. Peter Lang.

Pattison, J. (2020). From defence to offence: The ethics of private cybersecurity. *European Journal of International Security*, 5(2), 233–254.

Plaskota, L. (1996). *Noisy Information and Computational Complexity*. Cambridge University Press.

Rammert, W. (2012). Distributed Agency and Advanced Technology Or: How to Analyze Constellations of Collective Inter-Agency. In J.-H. Passoth, B. Peuker, & M. Schillmeier (Eds.), *Agency without Actors?: New Approaches to Collective Action*. Routledge.

Reviewing the Federal Cybersecurity Mission (2009) *Hearing before the Subcommittee on Emerging Threats, Cybersecurity, and Science and Technology, of the Committee on Homeland Security* (Serial No.111-5), U.S. House of Representatives, 111th Cong.

Rid, T. (2016). *Rise of the Machines: A Cybernetic History*. W W NORTON & CO INC.

Rid, T., & Buchanan, B. (2015). Attributing Cyber Attacks. *Journal of Strategic Studies*, 38(1–2), 4–37.

Rivera, J., & Hare, F. (2014). The Deployment of Attribution Agnostic Cyberdefense Constructs and Internally Based Cyberthreat Countermeasures. 2014 6th International Conference on Cyber Conflict (CyCon 2014), 99–116. https://doi.org/10.1109/CYCON.2014.6916398

Rowe, N.C. (2017). Ethics and Policies for Cyber Operations. In L. Glorioso & M. Taddeo (Eds.), *Challenges of Civilian Distinction in Cyberwarfare* (pp. 33–48). Cham: Springer.

Sandvig, C., Hamilton, K., Karahalios, K., & Langbort, C. (2016). Automation, Algorithms, and Politics: When the Algorithm Itself is a Racist: Diagnosing Ethical Harm in the Basic Components of Software. *International Journal of Communication*, *10*(0), 19.

Schandorf, M., & Karatzogianni, A. (2018). Agency in a Posthuman IR: Solving the Problem of Technosocially Mediated Agency. In E. Cudworth, S. Hobden, & E. Kavalski (Eds.), *Posthuman Dialogues in International Relations* (pp. 89–108). Routledge.

Schutte, S. (2012). Cooperation Beats Deterrence in Cyberwar. *Peace Economics, Peace Science and Public Policy*, *18*(3). https://doi.org/10.1515/peps-2012-0006

Singer, P.W., & Friedman, A. (2014). *Cybersecurity and Cyberwar: What Everyone Needs to Know*. Oxford University Press.

Skoudis, E., & Zeltser, L. (2004). *Malware: Fighting Malicious Code*. Prentice Hall Professional.

SkyQuest. (2024, July). *AI In Cybersecurity Market Size, Share & Forecast to 2031*. https://www.skyquestt.com/report/global-ai-in-cybersecurity-market

Slupska, J. (2019). Safe at Home: Towards a Feminist Critique of Cybersecurity. *St Antony's International Review*, *15*(1), 83–100.

Standish, R.K. (2001). On Complexity and Emergence. *arXiv:Nlin/0101006*, *9*. http://arxiv.org/abs/nlin/0101006

Stevens, T. (2020). Knowledge in the Grey Zone: Ai and Cybersecurity. *Digital War*. https://doi.org/10.1057/s42984-020-00007-w

Taddeo, M., McCutcheon, T., & Floridi, L. (2019). Trusting Artificial Intelligence in Cybersecurity Is a Double-Edged Sword. *Nature Machine Intelligence*, *1*(12), 557–560.

The White House. (2023). *National Cybersecurity Strategy*. https://www.whitehouse.gov/wp-content/uploads/2023/03/National-Cybersecurity-Strategy-2023.pdf

Thrift, N., & French, S. (2002). The Automatic Production of Space. *Transactions of the Institute of British Geographers*, *27*(3), 309–335.

Waldrop, M.M. (1993). *Complexity: The Emerging Science at the Edge of Order and Chaos*. Simon and Schuster.

Warren, K., Franklin, C., & Streeter, C.L. (1998). New Directions in Systems Theory: Chaos and Complexity. *Social Work*, *43*(4), 357–372. https://doi.org/10.1093/sw/43.4.357

Willson, M. (2018). Algorithms (and the) everyday. In D. Beer (Ed.), *The Social Power of Algorithms* (pp. 137–150). Routledge.

6 The Complex (Non-)Physicality of Information

The Existential, the Mundane, and the Logic of Noise

Introduction

"Cyberspace" has been frequently regarded as a "unique" space given its virtual nature. As a "virtual" space, cyberspace arguably transcends many of the constraints of the material or physical world. Communications that travel across borders in no time, virtual memories, virtual reality, virtual books, online shopping, and several other forms of virtualisation of life all make arguments of "immateriality" seem straightforward and intuitive. In fact, since the massive development of ICTs in the 1990s, many voices started to highlight the transformative influences of information by emphasising its immaterial nature (Mihalache, 2002). For many scholars, cyberspace represented "the substitution of bits for atoms" and an information society that "dematerialised nature" and transcended material objects by presenting on-screen equivalents (Dourish, 2017). Further, the so-called "cyber-weapons," which are primarily codes/software, are also sometimes portrayed as peculiar because of their purportedly non-physical nature. As argued by Dipert, "Cyberattack technology is more like an idea than like a physical thing" (Dipert, 2010, p. 404).

Although this immateriality cannot be denied, there is much more to information processes and operations than the immaterial and the virtual. That is to say, cybersecurity is peculiar, not simply as a virtual or immaterial sector, but because it embodies a complex relationship between the virtual *and* the physical as an intrinsic characteristic of information. This chapter, therefore, analyses the intersection between the physical and the virtual layer of information systems and how both layers co-constitute the logics of security in cybersecurity. The chapter focusses specifically on the logic of *existentiality* that has been regarded by many security literatures as a defining logic of security. Such literature, especially proponents of the Copenhagen School's securitisation theory, tie security to the existence of existential threats that threaten the survival of a particular referent object. This chapter, by contrast, argues that the peculiar (non-)physicality of information reduces the question of survival to be just *another* discourse in the construction of cyber threats, rather than a defining

DOI: 10.4324/9781003454113-6

logic of security. The (non-)physicality of information, the chapter contends, adds another important logic that contributes to the construction of *urgency without existentiality* in cybersecurity, which the book calls "the logic of noise."

Noise is a key concept in information theory and one important definition of entropy. The central aim of information theory is to maximise the amount of information in communication by minimising entropy – defined as noise – in the transmission channel or medium. As a problem of communication, entropy as noise is not viewed in existential terms in information theory. It is rather approached as *mundane* and *routine* disruption that tampers with information, but does not necessarily destroy it. Building on this understanding, the chapter presents the logic of noise as an important security logic in cybersecurity that evokes urgency without existentiality, and therefore highlights the significance of mundane cyber threats as opposed to the existential. Such logic, as will be explained in the chapter, is co-constituted by the simultaneous physicality and non-physicality of information.

To corroborate these arguments, the chapter proceeds in three sections. The first section summarises the philosophical debate in information theory and philosophy of information on the (non-)physicality of information. It then moves to an analysis of the complex relationship between the physical and non-physical elements of information systems. On one hand, it interrogates how the affordance of the physical infrastructure of digital information is determined by the intangible codes/software. On the other hand, it discusses the various physical articulations of the intangible elements of information representation and software operation and their constant interaction with digital matter. The second section of this chapter investigates the co-constitutive influences of the (non-)physicality of information in relation to general geopolitical security considerations. Here, it focusses particularly on the geolocation of data centres, data routing, cables construction, and software and hardware manufacturing as case studies. The third section moves from general security considerations towards a specific focus on the logic of existentiality and its implication in cybersecurity. It explains why existentiality in cybersecurity as an infosphere is not a precondition for perceived urgency, and develops a new theorisation of cyber threats through the logic of noise.

Information, matter, and the (non-)physical

The relationship between information and "physicality" has been subject to much philosophical debate. Many information theorists approach information materialistically or as a physical entity. They assume that information requires a medium to represent it – "no information without representation" – and that representation is necessarily tied to physical implementation. Written text exists as shapes on a paper or a screen, spoken words exist as acoustic waveforms, and in a technical context, they are stored as symbols in a computer or transferred through electromagnetic waves – which are all physical entities. Even ideas in a person's mind occur through neurons in their brains. All such examples of physical medium supports information and its very existence (Battail, 2013). Consequently, since information cannot be "physically disembodied" from the medium, some scholars argue that information is a physical entity per se (For example: Landauer, 1991, 1999).

The assumptions that "information is physical" and subsequently that "computers are physical systems" (Lloyd, 2000) are strongly connected to the emergence of information physics in the late 20th century. Building on Shannon's theory and the need for minimising the amount of energy used in transmitting information, information physics dealt with information as a "physical quantity" or as a "measure of interaction between physical systems" (Fradkov, 2007, p. 6). It is a physical phenomenon because all information processes, such as storing, transmitting, or processing data, involve varying levels of energy consumption and transduction, and are constantly influenced by thermodynamics and the laws of physics. These laws define what is informationally possible for devices used in information processing and set the boundaries of their development.

Although modern electronic information systems are designed to consume less energy, all information processes generate heat at every stage of transmission, encoding, and decoding (Karnani et al., 2009). Even the erasure of bits of information generates heat. Thermodynamics also influence the development of modern computers and laptops. The smaller a computer gets, the smaller the size of its microprocessor, the more difficult it is for heat to be released from it (Lutz & Ciliberto, 2015). That is why, it is argued that information theory and thermodynamics complement each other in their search for the best ways to utilise information within the available resources and energy levels, and that physical materiality is essential to computation theory (Floridi, 2010).

Nonetheless, physical representation does not necessarily impose physicality as an ontological property on information. For some theorists, information is an abstract rather than a concrete or physical entity. The fact that it has to be written down, transferred, and processed by physical medium does not mean that the information being written, transferred, or processed is physical as such. Although these theorists do not necessarily conceptualise information as entirely non-physical, they distinguish between the ontology of the medium and that of abstract information (Timpson, 2013). They argue that although the existence of information requires a physical medium, it does not necessarily depend on it. It is true that information requires a medium to exist, but it can exist on *any* medium. In this view, information is characterised by its invariance to the physical medium that carries or represents it.

The fact that information needs to be physically inscribed in matter does not mean that information is itself a physical entity, since the properties of the medium cannot be considered properties of information per se. For example, information has properties that cannot be possessed by matter or physical objects, including that fact that it is sharable and does not lose any of its parts when copied, unlike matter. It is also characterised by its proliferation capacity: it can exist in multiple medium simultaneously without increasing in number

(Battail, 2013). Even the simple idea that information is transportable and can travel at the speed of light – or even faster as argued by quantum mechanics – could be used to argue against its physicality (Burgin, 2010). This position is summarised by one study that defined information as a "non-physical emergent of particular physical processes" (Lombardi & López, 2017, p. 53).

Applied to digital information, many scholars argue that bits have a specific type of materiality that does not resemble that of the figures, letters, or sounds that are encoded in them. This is what one study called "bare materiality" to describe how the materiality of bits is not directly experienced by humans nor represented to them; it is rather directed to digital machines and systems (Evens, 2015). The semiotic relation between bits and humans occurs only after bits are decoded back to represent the original figures encoded in them. Further, although those encoded figures have a time and space-bound materiality, bits exist in a more abstract domain. They are designed and implemented in a way that allows them to isolate themselves from physical variables, like temperature and vibration, and only focus on the relevant material properties like data storage. By doing this, bits are subordinating matter and material properties to their logic (Evens, 2015).

This paradox of the (non-)physicality of information is evident in everyday interaction with ICTs. The digital is made of binary codes, embedded in machines and invisible to human beings. Codes/software hide their complexity in user-friendly virtual interfaces, fuelling the perception of the digital as immaterial, which one author described as "digital mysticism" (Boomen, 2009). Although people can touch screens and keyboards, they cannot touch codes and data. Codes are not written to be read or understood by human users, but by other digital machines. To the average user, codes seem as placeless entities, detached from the physical world. The electromagnetic medium that carries information appears to users as a "stampable mass," or a formless entity that can carry any electronic signal regardless of its content. Unlike regular mass that people touch and have bodily experience with, people move in "cyberspace" without an actual physical experience of space (Eldred, 2013).

In short, it is clear that the answer to the question *"is information physical?"* is not an easy one, and essentially, it should not be. Picking one side of the debate would result in reductionist view to the complex ontology of information, specifically in its digital form. Instead, this chapter argues that the property that makes information so peculiar is its *simultaneous* physicality and non-physicality. Unlike many of the aforementioned contributions, this chapter does not use the binary division between hardware as physical/material and codes/software as non-physical/immaterial in the analysis. Rather, it investigates how hardware, which is the obvious physical layer of cyberspace, is given meaning and functions by codes/software; and in the meantime, the chapter explores the way codes/software, as the obvious non-physical layer, have their own material or physical representations.

Digital information infrastructure

The materiality of information infrastructure is a very direct form of materialism. This infrastructure is usually referred to as the physical layer of cyberspace. In the computing and networking technologies, there are several manifestations of this materiality in different types of hardware, starting from the computer itself and its material parts, such as the central processing unit (CPU), hard disk, screen, keyboard, etc. There are many physical considerations that affect their operation, most importantly degradation. All these objects have a physical lifetime and specific capacities that can degrade over time or become technologically obsolete (Harvey & Weatherburn, 2018). Moreover, for networks to function they also require computer *servers*, which are high-end computers that perform functions for other computers (the clients) on a network. Networks also require internetworking devices such as *hubs* and *switches* that connect multiple client computers and allow them to communicate and share data. Similarly, a *router* (gateway) connects one network to another by routing data packets between them (Hallberg, 2009).

Another fundamental material part of the networking infrastructure are *cables*; usually considered as the backbone of networking. There are different types of cables; the most common of which are copper cables, also known as "twisted pair." Those cables contain multiple copper wires twisted together in a plastic insulator. When computers communicate in binary digits of zeros and ones, the wires transmit data as an electric current by changing voltages between two ranges, and thus the receiving system translates those two ranges into zeros and ones. There are various types of cables that differ according to the speed by which they transfer data and how resistant they are to outside interreference. The most expensive and sophisticated type of network cables are fiber-optic cables. Those are made of extremely thin and tiny glass tubes that use pulses of light rather than electric voltages to transmit data. Since they do not require electricity, they are not subject to electrical interference (Evans & Schneider, 2008).

When a user searches something on Google, for instance, the signals take a long route on physical devices, including routers, switches, cables stations, and undersea cables until they reach Google's data centre, and take a similar route back to the user. So, despite the wireless experience at the user's end and the illusion of immateriality, it is in the last stages of transmission between their computer and their home router is the signal finally set free from the materialities of the grid (Starosielski, 2015b). For that reason, it is sometimes argued that digital communications are no more than "magnetic flux on a disk, electrical currents, photons in optical cable" (Straube, 2017, p. 159). Some even believe that codes are no more than "signifiers of voltage differences," which means that basically, "there is no software" (Kittler, 1995).

Nevertheless, this physical infrastructure is not disembodied from codes/ software. For any object to become computational, it has to include computer codes to dictate the functions that it should perform. In digital technologies, codes/software are the forces that allow the physical to work by defining what it can possibly do and enable it to overcome its physical limitations. Everything that happens in the physical is a result of encoded bits. That is, the affordance of computational objects, or digital *matter*, is defined by codes and protocols that in themselves are not physical or tangible (Berry, 2011). These codes do not necessarily exist within the physical digital objects, they can instead interact with it externally. For instance, a DVD or credit card has no embedded codes, but if they do not interact with software to give them meaning, they will remain as mere plastic (Kitchin & Dodge, 2011). Berry argues, therefore, that code is a "super medium," or the element of information systems that puts all its part in one whole unitary shape (Berry, 2011, p. 10).

To conclude, the components of information systems infrastructure are material, yet *differently material*. They combine both physical materials, such as glass, plastic, silicon, and also intangible bits and electronic voltages (Bratteteig, 2010). The indispensable interaction with the intangible is what makes digital matter peculiar and different from matter in other security sectors. Digital matter is inherently informational, or 'information all the way down' (Dembski, 2016). However, this materiality of infrastructure does not just exist, it is also co-constitutive. It is obvious how a degrading hardware or broken cables can impact information operations. Most importantly, however, the material has the power to enable and/or constrain the virtual and its non-physical articulations, as will be shown next.

Information representation and software operation

The materiality/physicality of bits is a debatable issue in the fields of media studies, software studies, and the philosophy of information. Their intangibility has led many scholars to view them as inherently immaterial, even if they have material *properties*. This arguable immateriality is claimed to be the defining feature of the "information age" and its metaphysical promises in popular thought about liberating humans from the constraints of matter. In this view, the peculiarity of digital information is linked to its ability to avoid the boundaries of degradation that characterise physical matter. By entering the information age, they argue, the world has transformed "from atoms to bits." Through what is called "the method of abstraction," bits are capable of transmitting across various media, regardless of their physical properties. Put differently, "bits are bits" notwithstanding the physical medium of storage or transmission (Blanchette, 2011). However, there is more to codes/software than being just an abstract artefact or a "technology without matter" (Kallinikos, 2012). Although codes/software are "born digital," they are still bound by the physical and have their own physical representations (Dourish, 2017). Codes/software are not simply an abstract "self-contained language" separated from the material world. Rather, they are constantly and dynamically engaging with the physical in diverse forms of materiality (Zhu & Knoespe, 2007).

One important example of the influences of digital matter on the non-physical aspects of information is the process of developing software and programming. Although computing infrastructure was designed to transcend differences in physical computational resources, the historical development of this physical infrastructure has always had an impact on what is technologically possible in developing software. The transformation from wired to wireless communications for instance, or from desktop storage to cloud computing, had massive impact on software development. Computation is not just a method of abstraction to transcend physical material differences, but also a continuous trade-off process to make the best out of the limited material resources of storage, power, and connectivity. Efficient abstractions require a consideration of the "politics of resource allocation" (Blanchette, 2011).

Programming is another aspect of software development where the impact of the physical can be detected. To begin with, the process of writing codes passes with various forms of materiality when they are written as text on paper, compiled in hardware, tested by humans, and distributed through physical medium (Berry, 2011). Codes also depend on the practitioners' experiences and the "vernacular meaning" they give to them based on their embodied, real-world experiences. In its simplest form, computer commands like "print" or "copy" have a connection with natural vocabularies that are rooted in our experience of the physical world (Zhu & Knoespe, 2007). The interaction of codes with natural language can also be seen in the constraints of syntax in programming, such as punctuations, since the slightest mistakes can cause various errors. Added to this is the limited vocabulary of microprocessors that requires conformity between software and hardware, that is why, for example, some software is designed only for Mac and cannot run on Windows and vice-versa (Berry, 2011).

It is not just programming and software development that are influenced by the physical; such influences can be also traced on the operational level. Arguably, bits are not bits regardless of the media that stores and transmits them as Shannon's theory of information proposed. Bits are constantly communicating with the material, not just in the obvious implications of hardware design, but also software, and thus, they can be considered as "material objects" per se (Dourish, 2016). Although, theoretically, computers can only do what a program tells them to do, there are still other important material aspects in the execution of any task that programs do not specify. These include the network type, the computer's processing speed and memory size, the program's size, and the required memory capacity to execute it. These material characteristics influence the program execution and information representation and can even cause the program to stop working. The gap between the annotation in a program and the actual execution is where the materialities of information can be found, or where the "lie of virtuality" exists (Dourish, 2017).

Another manifestation of the deep connection between information representation and the material world is the problem of trade-off. For example, computers are supposedly capable of eliminating noise by using error-correction codes and thus mitigating physical constraints such as the network's bandwidth. However, error-correction codes increase data expansion and processing load, and in turn reduces capacity. Therefore, error-correction becomes a trade-off problem that involves various material constraints. Another example in networking is packet switching. Data travels throughout the network by being divided into packets that compete for limited processing bandwidth and thus impact applications like voice, streaming, and video by causing latency. These are all materially shaped trade-off processes (Blanchette, 2011).

Even in applications like emulation that is seen as "doubling the virtuality" of the virtual space, various materialisation can be pinpointed. In computing technologies, emulation is the way through which software is used in new devices to simulate older ones. For example, if a certain software used to work on an older version of a device and is no longer supported by the newer versions, emulation can create a simulated platform of the old device so that the software runs on the new hardware. This is similar to other forms of virtualisation, such as virtual books, virtual memory, and virtual reality. Although this may give an image of a completely virtual setting, there are multiple materialities that control it. In fact, rather than virtualising the old device, emulation actually rematerialises it using a host platform to bring it into action, and therefore remains constrained by the material properties of the present host and the absent emulated device. In this case, the differences in processors and capacities can obstruct the functioning of certain instructions in a program, resulting in errors or any form of performance reduction (Dourish, 2017).

Consequently, dealing with software as necessarily immaterial, or portraying digitality and materiality as two opposing categories, becomes obsolete. Such an argument overlooks the various material considerations that affect "virtual" representations and operations. Analytically, it thus makes more sense not to use the term "digital" to refer only to the non-physical or the physical components of cyberspace. Instead, "digital information" should be used as an overarching term that showcases the complex (non-)physicality of digital artefacts, whether in the form of physical infrastructure, or intangible codes/software. It follows that what makes digital information *different* is not the immateriality of its intangible elements, but rather the peculiar interactions between its tangible and intangible ones.

The geopolitical contexts of information systems: sovereignty, privacy, and security

The co-constitutive influences of the (non-)physicality of information on security can be further observed in studying the geopolitical contexts in which digital information systems operate. Such contexts create peculiar questions about sovereignty, privacy, and security that users are often entangled in, with little choice from their side. As will be shown next, every action that human users take on the internet can have various security implications connected to the geopolitics of information operation, that users are mostly unaware of. This unawareness extends to the location where their data is stored, the routes that their data packets take over the internet, the geolocation of the cables that carry them, and the manufacturing origin of the devices and software they are using.

On one side, digital information transcends the physical limitations of territories and distances. Through ICTs, one can cross borders by sending and receiving information from almost anywhere in the world. Geographical distance and proximity do not have to affect the speed or quality of the transmitted information. Besides, digital information is peculiar in its ability to exist in multiple places simultaneously. This makes geographical location as a means of identifying data ownership obsolete. As stated by a data expert and outlined by one study: "Sometimes the answer to the question 'where's my email?' is more quantum than Newtonian" (Blum, 2012, p. 240) – a statement that points out the fact that a single piece of information can exist in multiple places at the same time.

For instance, to speed up the process of retrieving data on the internet, some data may be replicated and stored in what is called "edge caches," which exist in closer locations to the user. This facilitates content retrieval by shortening the distance between the user and the server. The decision on what data is most in-demand and needs to be stored in cache is one that the "cache network" strategically and autonomously takes. In addition, in order not to waste resources and increase efficiency, data may be replicated in multiple servers in different regions. Copying data in several locations is also done to account for emergencies, like natural disasters, technical failure, or accidental loss that might affect any single data centre (Reisman, 2017). Another well-known practice in data storage as part of the cloud architecture is called "sharding." In this process, data stored in the cloud is divided into tiny portions, or shards, each stored in a different location across different regions (Hill & Noyes, 2019).

On the other hand, however, this purportedly "space-less," "immaterial," or "transcendental" experiences of digital information is made possible by massive physical infrastructure, with the influence of various geopolitical realities. These elements of materiality engender numerous security considerations that users mostly neither choose nor are aware of. In the following points, this argument is unpacked by analysing the geolocation of cloud data centres, data routing, cables construction, and software and hardware manufacturing. This analysis aims to support the two arguments that have been brought forward so far in this chapter: that information is simultaneously physical and nonphysical, and that this (non-)physicality is co-constitutive of peculiar security questions.

Data centres

Data centres are the core of cloud computing. Although cloud computing gives the user an impression that their data is stored in a virtual place, it is hugely physical. All the data in the cloud are actually stored in data centres and servers that have physical existence and that require a huge amount of energy to be run and cooled down.[1] A data centre is a facility that stores all the components of a computer system and its storage in a particular entity. They are a collection of cables, computers, routers, pipes, wires, hard drives, etc. These centres are not just material because of their physical representation, but also for the various geopolitical considerations they produce. When a user or entity in a certain country store their data in a cloud, or use it in accessing their emails, the data is not stored in a virtual, parallel space. Rather, it is stored in physical data centres that exist within certain territories that cloud providers choose based on IT costs, taxation policies, energy prices, etc. (Albeshri et al., 2014).

The management of data centres engenders various security concerns over data privacy and state sovereignty. Having citizens' data stored outside the borders of the state raises questions about legal jurisdiction over this data and the extraterritoriality of that state's laws and regulations. A famous example here is the legal battle between the USA and Microsoft in 2014, when the government demanded access to the emails of an American citizen that were stored on Microsoft's data centre in Ireland (Daskal, 2018).[2] Had the data centres been in the USA, this legal controversy would not have taken place because the centres would be under the US jurisdiction. Likewise, if the US government were able to force Microsoft to hand in the data it requested, it would have been a breach of the Irish data privacy. As said by a witness from Microsoft in a congressional hearing, explaining why non-American cloud users could be discouraged from using their services:

> So I should use a local provider, right? Because if I use your cloud service, you are a global company; you are headquartered in the United States. You are just going to give all our data to the U.S. Government.
> (Protecting America from Cyber Attacks, 2015, p. 24)

For these reasons, many states perform data localisation or data territorialisation practices. This is done in the form of regulations demanding the storage of citizens' data on data centres located within their borders (Baur-Ahrens, 2017). Many forms of data localisation laws are already implemented in several countries with varying degrees, including Brazil, Russia, India, China, among others (Selby, 2017). Also, following the Snowden revelations, several countries in Europe started calling for a "European cloud infrastructure." In Germany, for instance, Deutsche Telekom – its biggest telecommunication corporation – called for storing all citizens' emails locally in a campaign titled "E-mail made in Germany" (Baur-Ahrens, 2017). Yet, data localisation can also have negative implications on citizens' privacy and their personal data security. The local storage of data may be a barrier towards the implementation of international security standards for data protection, which global companies have to abide by given the competitive environments they operate in (Fraser, 2016).

Data routing

The internet infrastructure is not fixed and linear as telephone systems; it is composed of a wide range of scattered, non-hierarchal nodes and hubs. Instead of relying on a centralised entity, the functionality of communications on the internet is managed by self-organising end hosts. This flexibility is meant to secure communications against disruptions resulting from targeting a central hub (Baur-Ahrens, 2017). When data transfers over a network, it is divided into packets, and each packet takes a certain route, until they are re-assembled at their destination. The path the packets take is not a decision that the user make, nor is the user mostly aware of. Rather, it is decided by the router itself according to the distance between the source and destination, bandwidth, number of hops on the network, and several other factors to ensure the efficiency of delivery (Misra & Goswami, 2017).

Routing is an aspect of the internet architecture that is highly influenced by several geopolitical considerations. The fact that the network traffic inside a country may leave its borders even if the sender and receiver are based locally raises questions of security, sovereignty, and privacy. That is why, some countries have called for having a "national internet" or "domestic internet," by trying to localise data routing that takes place on their territories. China and Russia are among the countries that implement certain aspects of national routing. Additionally, in 2013, Deutsche Telekom also campaigned for a "German internet," alongside the previously mentioned campaign for localising data storage. But since these attempts were not conforming with European laws, the campaign shifted to calling for "Schengen routing" or making sure that communications sent within the Schengen area do not get transferred through foreign territories (Heumann, 2017).

Undersea fiber-optic cables

Although the current age is marked by increased "wirelessnes" among a wide range of devices, these wireless connections are supported by a huge infrastructure of cable systems, under soil or under sea. Fiber-optic cables in specific appeared to be more cost-efficient and better in capacity than satellites since the 1990s (Starosielski, 2015b). Right now, most of the internet that travels across the ocean is transmitted via cables, not satellites. The undersea fiber-optic cables are responsible for transporting 99% of digital communications across the ocean; that is why they are considered the backbone of the internet (Chesnoy, 2015). And given the decentralisation of data routing, even if the sender and receiver of the data exist within the same state, the data packets might be transferred through cables outside its borders.

The geolocation of cables raises similar security and privacy concerns to routing and data centres, since their location can make the data passing through them susceptible to surveillance. The documents released by Edward Snowden revealed that the NSA and the GCHQ (the British intelligence organisation) were wiretapping the data flowing in fiber-optic cables between Google and Yahoo data centres, as part of a project they called "MUSCULAR" (Gellman & Soltani, 2013; Rushe et al., 2013). As a result, several new cable projects in Europe, Asia, and the Middle East were proposed to route networks away from cables located in the USA (Starosielski, 2015a). However, the challenge remains that the distribution of cable infrastructure is both centralised and limited in terms of available paths. For instance, currently, there are more than 400 undersea cables, around 88 of which connect the USA to the world (Long, 2023). This is considered a big number compared to other countries that have five or less external links. This concentration can be explained by the low financial incentives to diversify location, in addition to security considerations linked to scarcity of safe location for cable extensions (Starosielski, 2015b).

Most recently, subsea cables have become an area of superpower competition, especially between the USA and China. In February 2023, SubCom LLC, an American company, started a project for a $600 million cable, known as South East Asia-Middle East-Western Europe 6, or SeaMeWe-6, which transports data between Europe and Asia via the Middle East and Africa. The contract for this project was competed for by various Chinese companies, but it has been reported that the US government intervened and succeeded in changing the route of this contract, out of fears from potential spying by the Chinese government (Brock, 2023). This is also the reason why the US government denied Google, Meta, and Amazon permission to establish subsea cables that connect the USA and Hong Kong, fearing that this would facilitate spying by the Chinese government. In retaliation, China is also delaying the approval of cable projects, thus causing huge financial losses to companies (Braw, 2023).

Hardware and software manufacturing

Hardware and software are not merely technical products detached from their socio-political context; and one key aspect of this context is geopolitical. Where software and hardware are produced, and the nationality of the companies that manufacture them, is an important cybersecurity consideration. As mentioned in the 2011 defence strategy for operating in cyberspace by the DoD, ICTs products are manufactured and assembled in different places, and can be maliciously tampered "at points of design, manufacture, service, distribution, and disposal" (The Department of Defense, 2011, p. 3). This has been an issue of concern to the US government for a very long time. For example, in 2003, the NSA information assurance director stated that the USA should manufacture the software used in CNIs locally, in order not to risk it being compromised by foreign nations (Cybersecurity – Getting It Right, 2003, p. 22).

Several cases demonstrate the criticality of this issue. For instance, the leaked Snowden files revealed that the NSA and GCHQ exerted pressure on private companies for surveillance purposes, including Microsoft, Apple, Facebook, Google, among others. This was done through court orders, withholding licences, or hacking into their systems (Deibert, 2015). These measures altered the software or hardware of targets' devices, a process they called "interdiction" (Biham et al., 2016); weakened encryption by utilising supercomputers capable of cracking encryption algorithms; and enforced "backdoor" access to software (Harding, 2014). This is similar to the backlash faced by Huawei and ZTE, forcing them to exit the US market, amidst fears by the US government that the Chinese government might be embedding backdoors in their products or have access to their data (Bartz et al., 2022).[3] Most recently, the Biden administration announced the intention to ban the sales of Kaspersky, the antivirus software produced by the Russian company Kaspersky Lab, arguing that it threatens national security (Alper, 2024). In fact, even before this decision, the Russian origin of the company has always created similar concerns. For example, in the past, the DHS took the decision to stop using Kaspersky Lab and ordered all government agencies to follow suit, claiming that it is linked to the Russian intelligence (Rosenberg & Nixon, 2017). The same also applies to the app TikTok, which has been banned in various countries, either for government entities or the general public, because of fears of its ties with the Chinese government, including in Australia, Estonia, the UK, France, the Netherlands, India, Pakistan, and several other countries (Euronews, 2024).

This analysis shows that the materiality of the geographical context in which information flows matters for security. This is particularly important given the imbalance in the distribution and control over the physical infrastructure mentioned earlier. For example, data shows that around half of a 2.5 billion analysed internet traffic goes through at least one member of the Five Eyes intelligence alliance: the USA, the UK, Australia, Canada, and New Zealand. This means that with the right technology, those countries have the technical capacity to spy on a huge number of data. In addition, the majority of technology companies that most people around the world use, like Google, Apple, and Facebook, are all based in the USA, which makes them compliant to the US law (Buchanan, 2020). Although information has the capacity to break the boundaries of physical matter, therefore, it still remains simultaneously physical. This peculiar (non-)physicality, moreover, is co-constitutive of cybersecurity logic(s). Specifically, this property of information poses challenges to the logic of existentiality in security theories and when applied to cybersecurity – this will be further explained in the next section.

The (non-)physical between existentiality and noise

Existentiality holds a central position in many security studies literature. A prominent example in this regard is the Copenhagen School's securitisation theory which has been extensively applied to the study of cybersecurity, as explained in previous chapters (Bendrath, Eriksson, & Giacomello, 2007; Dunn Cavelty, 2008a, 2008b; Hansen & Nissenbaum, 2009; Kallender & Hughes, 2017; Lacy & Prince, 2018; Aljunied, 2019; Hassib & Alnemr, 2021). According to this body of literature, existentiality is integral to the conceptualisation of security and is considered an indispensable quality of the threats that security aims to survive. Existentiality is also intrinsic to the conceptualisation of the referent objects of security. The securitisation theory defines referent objects as "things that are seen to be existentially threatened and that have a legitimate claim to survival" (Buzan et al., 1998, p. 36). What cannot be existentially threatened, or perceived as such, cannot be considered a referent object of security. So, for example, firms in a liberal economy cannot be considered referent objects since they are not expected to last forever and therefore cannot securitise their survival (Buzan et al., 1998).

Many literatures in CSS contend that security is always about the existential and the exceptional and thus that it must be criticised and rejected. Even scholars like Floyd, who did not necessarily reject security but argued instead for a "just securitisation theory," accepted existentiality as a given. Floyd asserted that the "moral rightness of securitisation" will be achieved if threats are "objectively existential," by establishing causal relations with a particular intentional aggressor. Accordingly, immigration, for instance, cannot be constructed as an existential threat because the intentionality of harm by a particular aggressor is not achieved (Floyd, 2011, 2015).

Whether digital information can be existentially threatened or constructed as such is an important question that the theoretical literature on cybersecurity did not sufficiently engage with. Most of these literatures focussed on studying cyber threat representations, without examining whether they match any criteria for existentiality and whether existentiality is in itself an applicable logic to cybersecurity to begin with. Instead, many studies focussed on conceptualising "extraordinary measures" and whether they have been applied to cyber securitisations. Further, many literatures implicitly assume that it is enough for a threat to be presented as *serious* to qualify as *existential*. This is evident in many cyber securitisation literatures, partially due to their preoccupation with state and political discourses, in which claims of existentiality are usually made. Most importantly, it can also be explained by their focus on human discourses without considering the role of the non-human referent object in shaping/limiting what is possible in constructing cybersecurity.

Acknowledging the co-constitutive role of information and the influences of its (non-)physicality shows that existentiality is just *another* discourse in cybersecurity. It is not the only reason for threats to register in the cybersecurity debate, nor is it a precondition for perceived *urgency*. That is to say, existentiality and questions of survival are not intrinsic to cybersecurity and are not essential legitimisers to urgency and immediacy. This chapter introduces another important logic that helps to understand and theorise the complex construction of cyber threats, which is that of "noise." As explained earlier, noise as disruption or interference in signal transmission lies at the core of Shannon's information theory, as one key definition of entropy. Shannon regarded noise as a key problem in information communication, albeit one that is not existential in nature.

To build the foundation of noise as an alternative security logic to existentiality, we need to first interrogate the meaning of existentiality as such. Wæver, a prominent security scholar and one of the Copenhagen school's theorists, contends that an existential threat is one that targets the "essential being" of the referent object; not one that simply results in varying degrees of harms (Wæver, 2009). Constructing a threat as existential is like saying: "If we do not tackle this problem, everything else will be irrelevant (because we will not be here or will not be free to deal with it in our own way)" (Buzan et al., 1998, p. 24). Noise as a challenge in information theory, on the contrary, does not meet this criterion of existentiality. In information theory, noise is portrayed negatively as the "parasite" of communication, yet one whose existence does not threaten the *survival* of information. It is a threat to information and communication, though not an *existential* one. Shannon's theory aimed at maximising the amount of information in a transmission channel *despite* the existence of noise, which means that information and noise exist simultaneously. To develop "noise tolerance" for information transmission, Shannon introduced methods such as redundancy and error-correction (Fresco & Wolf, 2016). Likewise, coding theories tend to encode information in the transformation process in such a way that allows the retrieval process to take place despite noise (Piccinini & Scarantino, 2016, p. 27).

That is, noise is *less than* existential. It is not a destructive phenomenon in communication channels (Krapp, 2011). Information does exist despite the disruption caused by noise. Moreover, even when viewed negatively as a problem for communication, noise remains integral to the existence of information. As put by Malaspina, "…the creation of information can only occur on the basis of noise" (Malaspina, 2018, p. 75). As such, noise is sometimes considered as a precondition for complexity and a reflection of the variety of a system. Hence, it is an essential concept in complexity theory and computer science, in which the idea of creating order out of entropy – defined as noise or disorder – is explored. Minimising noise and contingency are key goals in the operation of information systems, but the existence of both does not in itself threaten the existence of information.

Noise can be used analogically as a logic of security to help understand the complexity of cyber threats, as opposed to the centrality of the existential in many theoretical cybersecurity literature. The majority of cyber threats in the documents analysed in this book are viewed as *disruptive* rather than destructive. This marks a belief that the cyber threats we should be concerned about are not necessarily the ones that disable the target, but rather the ones that manipulate it "in a very unintended fashion" (Securing Critical Infrastructure in the Age of Stuxnet, 2010, p. 45). When disruption is portrayed as the consequence of a cyber threat, several referent objects that intersect with other security sectors are drawn into the discourse. Links to the economic sector are established when cyber threats are seen as harmful to "economic competitiveness," "business opportunities," "innovation," "customers' confidence," the state's "global competitive advantage or leadership," etc. Links with military security are found when the cyber threat is viewed as a challenge, not to the survivability of armed forces per se, but to their operations and communications, defence and emergency capabilities, and their ability to use cyberspace as a force-multiplier. Added to that, cyber threats are sometimes portrayed as a security challenge for "internet openness" or "cyberspace openness," which in turn affect the privacy and civil liberties of individuals. All such threats are mainly interpreted in disruptive rather than destructive terms.

Just like noise is perceived as an intrinsic part of information systems, but one that has to be battled, the disruptive implications of cyber threats are viewed negatively yet not existentially. They are often characterised as "'catastrophic," "debilitating," "massive," "critical" to the economy and the society, even if they do not necessarily threaten their survival. Here, urgency and immediacy in constructing the cyber threat is often built on an assumption that cyber technologies are growing more complex, and with complexity comes more insecurities and greater risks, because "Complexity is something we can't change" (Overview of the Cyber Problem, 2003, p. 11). Complexity is perceived as both a defining feature of the technology and of its attack tools, including malwares. Increasing complexity and dependency widen the attack surface and render the implications of a cyber-attack more severe. This is seen as one factor that contributes to shifting the offence-defence balance towards the offence advantage. Additionally, urgency is evoked when past cyber-attacks are mentioned, with the acknowledgement that they had disruptive rather than destructive ramifications; mainly financial losses and operational dysfunctions.

This does not mean, however, that cybersecurity discourses are not full of futuristic disaster scenarios about potential destruction. Although advocated mainly by the intelligence community, analogies of "cyber Pearl Harbour" and "cyber 9/11" are widely adopted by many other actors who present the cyber threat in destructive terms, even if not part of the official strategy of the state (Lawson, 2019). For example, The NSA director's statement that a cyber Pearl Harbor "is not a question of if but when" is highly referenced in many statements by MPs, security experts, and private corporations (America Is Under Cyber Attack: Why Urgent Action Is Needed, 2012, p. 46). Yet, discourses of cyber destruction are not dominant, and are usually marked by wide uncertainties: the destruction is often perceived as *potential* or *possible* but not *certain*. It is primarily in the defence and intelligence community discourses that more assertions are used. But if the analysis is widened to include private actors, the story becomes totally different.

Here, the analogy of noise is relevant in recognising the significance of *mundane cybersecurity*. There is no doubt that some cybersecurity discourses match the existentiality assumption. High-profile cyber incidents that are widely publicised, such as Stuxnet, WannaCry, Notpetya, and others, are sometimes used as a basis for an argument about survival. However, the cybersecurity challenge cannot be reduced to the threat of one big incident, crises, or disaster that subscribes to an existentiality assumption. Unlike the scenarios long imagined by many academics and cyber strategists talking about the potentiality of cyber wars that resemble that of a nuclear catastrophe, none of that has actually taken place. Instead, cyber incidents are becoming the 'new normal of geopolitics'; i.e., as mundane as noise in the operation of information systems. They are happening on daily basis in a persistent, albeit non-destructive manner. They destabilise world politics, without needing to be apocalyptic (Buchanan, 2020).

In fact, the majority of cyber-attacks that are seen as the most serious in history were neither objectively existential from a technical viewpoint, nor portrayed as such by the majority of concerned actors. This does not mean that the cyber threat is not sometimes hyped or exaggerated, since these two qualities are not essentially linked to existentiality and survival. But why so? Explaining why and to what extent existentiality may or may not register in the cyber threat perception should not be reduced to the thoughts, interests, and intentions of cybersecurity actors. There is much more that the properties of information can say about existentiality in cybersecurity, particularly its complex physicality and non-physicality.

The physical and the logic of existentiality

In debating whether information should be theorised as a physical or a non-physical entity, some information theorists resort to existentiality for an answer. They analyse the physicality/non-physicality of information by addressing a key philosophical question: can information be destroyed? For instance, some contend that the destruction of a physical media carrying information, such as books or hard desks, does not mean the destruction of information per se (Ben-Naim, 2008). This is used as an argument for the non-physicality of information since it exists in a "spatiotemporal organisation" of energy and mass that means it cannot be destroyed (Tse, 2013). Likewise, the non-physical aspects of digital information arguably condition the perception of existentiality and reduce it to the physical. That is, when the cyber threat is presented in existential terms, it is usually associated with the *physical* elements of information systems and attacks that could result in physical consequences. This goes against many security literature's assumption that survival is not necessarily tied to the concrete or physical, because in cybersecurity *it mostly is.*

Existential threats in cybersecurity are mostly connected to the fear of potential physical damages resulting from a cyber-attack. That is why, whenever the cyber danger is discursively aggravated, the physical is brought into the argument. This emphasis on the destructive physical implications of cyber-attacks is the closest to the existentiality assumption, particularly if the target is CNIs.[4] Given their interconnection with medical systems, power plants, and the emergency response capabilities of the state, cyber-attacks on CNIs are usually constructed as existential threats because they will necessarily result in *physical damage.* When attacking CNIs is discussed, it is usually accompanied by several scenarios of physical damage such as planes crashing, lethal clouds emitting from chemical plants, exploding pipelines, total national blackouts, etc. (Burton & Lain, 2020). Here, the existential threat is either the direct physical damage to the infrastructure, or the indirect one in the form of potential loss of life.

For example, perceptions of potential loss of life as a consequence of cyber threats is a foundation upon which the newly emerging field of *cyberbiosecurity* is established. This field, as introduced by scholarly literatures in the life sciences, addresses cyber risks resulting from digitising biology, including embedding malware in DNA, corrupting gene-sequencing, manipulating biomedical materials, stealing epidemiological data, or developing biological weapons in a way that threatens *human life* and survival (Fouad, 2024). This is closely connected too to the growing fears of cyber-attacks targeting health care. A famous example here is a cyber incident that disrupted the information system of a hospital in Germany, leading to the death of a patient who could not be admitted to the hospital as a result of this system outage (Reuters, 2020). However, the argument that cyber-attacks can cause loss of life here may not be very straightforward, due to the indirect nature of the consequences and the difficulty of establishing causation between the cyber-attacks and the patient's death.

Furthermore, not only do the physical elements of information systems allow for the existentiality assumption to feature in cybersecurity discourses, they also take away the existential quality from the non-physical. In such discourses – in which the physical damage resulting from a cyber-attack is considered possible in the future – identity theft, espionage, and attacks that lead to the disruption of services or loss of data are not seen as existential enough, since they are not physically destructive. This argument can be summarised in the following opening statement from the Chairman of the Subcommittee on Economic Security, Infrastructure Protection and Cybersecurity in the House of Representatives, Daniel E. Lungren:

> Many of us recognize the average cyberattack such as a worm or virus is a nuisance, one that irritates us, slows down our computers or prevents us from e-mailing. Yet deliberate cyberattacks have the potential to do physical harm in the form of attacks on cybersystems controlling critical infrastructures.
>
> (H.R. 285: Department of Homeland Security Cybersecurity
> Enhancement Act of 2005, 2005, p. 2)

Nonetheless, discourses that emphasise this logic of existentiality are still not dominant outside the scope of the military and the NSA and are primarily centred on the *potential* rather than the *certain*. One obvious reason for this is the fact that such attacks have not taken place before, or at least in the same scale mentioned in such existentiality-induced discourses. Most importantly, as argued by Rid, the violence resulting from a cyber-attack is inherently both *indirect* and *less physical* than conventional forms of violence. The tools of cyber-attack, commonly referred to as "cyber weapons," are generally not lethal and rely on "weaponising the target" and utilising its energy instead of inheriting the violent nature themselves. Put differently, codes are bound by their indirect nature as a medium of presumed violence and do not have an inherent "explosive charge" (Rid, 2013). The indirect nature of the majority of cyber-attacks and the non-physicality of their consequences challenge the existentiality logic.

For instance, it is widely believed by many actors in the USA that there are two types of entities in the modern time: those who know they are hacked and their security is compromised and those who do not – with particular reference to the Chinese and Russian governments as the hackers. This is not usually represented as an existential threat as such. Yet, if the same argument was to be made in the military sector, the claim of existentiality would be more straightforward. Moreover, there is also a belief that most states maintain a presence on other governments' networks for espionage purposes, in which malwares are used to breach the systems. Due to the immateriality of the informational targets and the tools used in those operations, the existentiality claim cannot be easily established. As put by Michael T. McCaul, a representative in Congress, in the context of discussing industrial espionage from China:

> You know, we talk about the analogy, agents of a foreign power caught with paper files walking out with classified or nonclassified information, it will be all over the papers. But yet in the virtual world, that is happening and no one seems to know or really pay attention to it.
> *(America Is Under Cyber Attack*, 2012, p. 45)

Hence, even when existentiality is in question in cybersecurity discourses, it is usually connected to high-profile hostile operations that are widely publicised in the media. However, cybersecurity is not just about these attacks. At the heart of cybersecurity lie the less-than high-profile, mundane incidents that take place on daily basis across the world, targeting a wide range of entities and individual users. These threats are closer to the logic of noise than existentiality. In turn, the anti-entropic policies implemented to defend against them – such as patching, intrusion detection, and the rest of practices mentioned in detail in Chapter 4 – resembles what Huysmans called "little security nothings" (Huysmans, 2011). This is a kind of security that is not centred on the exceptional, but one that extends to the banal, everyday, and routine practices.

The non-physical and the logic of noise

If the physicality of digital information co-produces existentiality perceptions, the non-physical elements limit them and open the door for noise instead. This can be explained by three main reasons. The first is the *invisibility of cyber insecurity and the absence of adequate imagery*. Whereas the physical bring the "cyber" closer to our "real-world" experiences and imagination of danger, the non-physical makes it hard to visualise the consequences of cyber threats through photographic imagery. In security studies, there is a growing body of research on the role of visuality in constructing security as a part of a field known as visual security studies (Vuori & Saugmann, 2018), together with a growing use of visual methods in CSS in general (Andersen et al., 2014).

Visual security studies illuminate many peculiar characteristics to visuality in securitisation that cannot be easily utilised in cybersecurity. For example, Hansen argues that images prompt instant emotive reactions that go far beyond the kind of reactions people have towards texts or words. Images of violence, for instance, can have powerful emotional impact on the observer that is incomparable to speaking about it (Hansen, 2011). In addition, images have a "special affinity to reality" and are capable of creating a sense of authenticity. This is driven from an assumption that what the camera captures is an objective reality. There is also a strong link between images and temporality. Images preserve memories. By simply looking at it, an image has the power to instantly connect the observer to particular historical phases or memories that evoke certain emotions. Some images may even turn into icons that have a specific interpretation in the viewers' minds (Brink, 2000).

However, visuality and imagery in cybersecurity are conditioned by the non-physical aspects of digital information. Cyberspace is "an invisible battle ground" (*Securing the Modern Electric Grid from Physical and Cyber Attacks*, 2009, p. 21). We cannot possibly visualise a phishing campaign or have a photographic imagery of the aftermath of data being stolen. As explained by the director of the National Cybersecurity and Communications Integration Centre in the DHS: "...if I told you there was a Category 4 hurricane that hit the Gulf Coast you would go, 'Oh, that is bad.' Category 1? It is bad, but 4 is worse....What is that in cyber? How do we get that imagery?" (Facilitating Cyber Threat Information Sharing and Partnering with the Private Sector to Protect Critical Infrastructure, 2013, p. 28). It is not just the absence of previous experience of cyber destruction that leads to a situation that "people are much more afraid of bombs and anthrax than they are of viruses and worms" (*H.R. 285: Department of Homeland Security Cybersecurity Enhancement Act of 2005*, 2005, p. 12). Rather, it is also the inherent invisibility of cyber insecurity. The fact that when a cyber-attack takes place "There are no burning buildings or collapsing structures" (Overview of the Cyber Problem, 2003, p. 6) makes existentiality hardly imaginable. This is also arguably one reason why people care less about their digital privacy than they would in a non-digital context. Digital surveillance is intangible, cannot be seen, and therefore hardly felt, which gives an illusion of privacy (Mitnick, 2019).

Secondly, information is innately replicable and therefore possibly retrievable. Unlike atoms, bits are persistent by default (Boyd, 2010). The non-physical aspects of bits makes it easy for digital data to be duplicated and copied and impossible to distinguish a copy from an original (Masur, 2018). In addition, as stated earlier, there is always a possibility that data exists in multiple places, so an attack on one does not mean necessarily its complete loss. If spied on and breached by cyber-attacks, like espionage, data can still remain intact and unharmed. Stealing military, commercial, or personal information does not necessarily affect the survival of the state, the private sector, any individual, or the stolen data itself. Similarly, denying customers/citizens access to certain services through DDoS attacks, for instance, does not in the majority of cases threaten the essential being of anyone. This does not mean, however, that data is never lost; because it can and does. But this replicability and retrievability property makes its complete annihilation in case of an attack *possible* rather than *inevitable*. That is why, it is often argued that the majority of data loss due to failure of software or hardware, human errors, or malware are recoverable (Reuvid, 2006).

Thirdly, the universality of computing devices makes them inter-changeable and can weaken the logic of existentiality. As stated earlier, one property of codes/software is that they are not tied to a particular matter. If a computer gets hacked, in most of the times the user can re-install a new operating system and still use it. If it happens to stop working due to this attack, they can re-install their backed-up data on a different device and still have the same experience. The same applies to software; their existentiality is hardly ever a question in the case of a cyber-attack. When a software is attacked, it remains operable after patching and does not cease to exist. It can be re-designed, re-written, re-tested, or debugged, but is not destroyed due to an attack. Though software is not eternal, it may cease to exist due to the development of new technologies that make it obsolete, its incompatibility with new devices, or several other reasons, but not as a direct result of a cyber-attack. Here, the cyber-attack is the noise that threatens the system but does not necessarily challenge its existence.

Conclusion

This chapter focussed on the third property of information that the book argues is co-constitutive of cyber (in)security, which is its simultaneous physicality and non-physicality. Against the seemingly intuitive arguments that "cyberspace" and digital information are essentially peculiar because of their "immateriality," the chapter showed that their peculiarity actually stems from a complex interaction between their *simultaneously* physical and non-physical nature. This (non-)physicality co-produces several security conditions that human actors may involuntarily be tangled in. It also co-constructs a different conceptualisation of existentiality and urgency in cybersecurity and opens the way towards an analogical interpretation of cyber threats through the logic of noise.

On one hand, the infrastructure of digital information is a clear signification for materiality. Information cannot exist without representation, and this representation is always through a physical medium. Computers and their components, hubs and switches, cables, and the rest of internetworking infrastructure is key for digital information processes. Yet, digital matter is not just *another* type of matter, because it is inherently informational. The physicality of digital matter is not disembodied from the intangible or the "virtual." Its affordance, functions, and capacities are dictated by codes/software. On the other hand, codes/software have their material articulations and are constantly interacting with the physical world. These interactions are manifested in the process of software development, programming, code-writing, and in software operation. Bits are not bits regardless of the medium as some scholars argue, and both their operation and the way they are experienced by the end-user are affected by the physical properties of the infrastructure.

The entanglement of the physical and non-physical is evident in the geopolitical context of digital information operation. Digital information is characterised by its divisibility and mobility and is peculiar in its ability to exist in multiple places at the same time. This does not mean, however, that "cyberspace" is placeless or borderless. The geolocation of data centres and undersea fiber-optic cables, the paths that data routing take, as well as the geographical origin of hardware and software manufacturing are all demonstrations of the argued (non-)physicality of digital information. This, in turn, creates various security considerations and privacy concerns that users mostly neither choose nor are aware of, with every simple action they make online.

In addition to the general security considerations regarding privacy and sovereignty, this (non-)physicality of digital information reduces existentiality to being just *another* discourse in the construction of cyber threats, that is limited to the physical. If not existential, the cyber threat is constructed as urgent and imminent through what the thesis calls the logic of noise. Viewed as disruption to communication channels in information theory, noise can be used analogically to understand the complexity of cyber threats in a space between contingency and control. The same as noise is seen as a "normal" characteristic of information operation, but one that needs to be minimised and challenged, cyber threats are often viewed as a disruptive yet integral aspect of the everyday functioning of systems. Reasons include the absence of an imagery for the non-physical, the innate replicability of digital data, the universality of computing devices, the invisibility of attacks and uncertainties about the scope of damages, and the indirect nature of cyber violence. It is the physical that can bring existentiality to the analysis, particularly in regard to the security of CNIs.

Notes

1 In some estimates, one data centre can use the same amount of power required for a medium-size town, or even more, making the cloud hold the fifth place in world electricity demand. And because of the energy waste they produce, many companies are now trying to apply less-energy intensive strategies, by relying on hydropower and renewable energy as part of "greening" cloud infrastructure (Vonderau & Holt, 2015).

2 This legal battle continued until the CLOUD Act was passed by the Congress and signed in 2018. The act allows the government to compel American technological companies to provide it with data it requests, even if it is stored on foreign soil, subject to data sharing agreements between the USA and foreign governments (Daskal, 2018).

3 In the same vein, a news report was published in October 2018 claiming that China inserted a backdoor in servers' chips used by around 30 US companies and

government entities, including Amazon and Apple, during the manufacturing process in China (Robertson & Riley, 2018).

4 In the cybersecurity strategy of 2003, critical national infrastructures were defined as the "public and private institutions in the sectors of agriculture, food, water, public health, emergency services, government, defense industrial base, information and telecommunications, energy, transportation, banking and finance, chemicals and hazardous materials, and postal and shipping" (The White House, 2003).

References

Albeshri, A., Boyd, C., & Nieto, J.G. (2014). Enhanced Geoproof: Improved Geographic Assurance for Data in the Cloud. *International Journal of Information Security, 13*(2), 191–198.

Aljunied, S.M.A. (2019). The Securitization of Cyberspace Governance in Singapore. *Asian Security, 0*(0), 1–20.

Alper, A. (2024, June 21). Biden bans US sales of Kaspersky software over Russia ties. *Reuters.* https://www.reuters.com/technology/biden-ban-us-sales-kaspersky-software-over-ties-russia-source-says-2024-06-20/

America is Under Cyber Attack: Why Urgent Action is Needed 2012 *Hearing before the Subcommittee on Oversight, Investigation, and Management, of the Committee on Homeland Security* (Serial No. 112-85), U.S. House of Representatives, 112th Cong.

Andersen, R.S., Vuori, J.A., & Mutlu, C.E. (2014). Visuality. In C. Aradau, J. Huysmans, A. Neal, & N. Voelkner (Eds.), *Critical Security Methods: New Frameworks for Analysis* (pp. 95–117). Routledge.

Bartz, D., Alper, A., & Bartz, D. (2022, December 1). U.S. bans new Huawei, ZTE equipment sales, citing national security risk. *Reuters.* https://www.reuters.com/business/media-telecom/us-fcc-bans-equipment-sales-imports-zte-huawei-over-national-security-risk-2022-11-25/

Battail, G. (2013). *Information and Life.* Springer Science & Business Media.

Baur-Ahrens, A. (2017). The Power of Cyberspace Centralisation: Analysing the Example of Data Territorialisation. In M. Leese & S. Wittendorp (Eds.), *Security/Mobility: Politics of Movement* (pp. 37–56). Oxford University Press.

Bendrath, R., Eriksson, J., & Giacomello, G. (2007). From 'Cyberterrorism' to 'Cyberwar', Back and Forth: How the United States Securitized Cyberspace. In J. Eriksson & G. Giacomello (Eds.), *International Relations and Security in the Digital Age* (pp. 57–82).

Ben-Naim, A. (2008). *A Farewell to Entropy: Statistical Thermodynamics Based on Information.* World Scientific.

Berry, D. (2011). *The Philosophy of Software: Code and Mediation in the Digital Age.* Springer.

Biham, E., Carmeli, Y., & Shamir, A. (2016). Bug Attacks. *Journal of Cryptology, 29*(4), 775–805.

Blanchette, J.-F. (2011). A Material History of Bits. *Journal of the American Society for Information Science and Technology, 62*(6), 1042–1057.

Blum, A. (2012). *Tubes: Behind the Scenes at the Internet.* Penguin UK.

Boomen, M. van den. (2009). *Digital Material: Tracing New Media in Everyday Life and Technology.* Amsterdam University Press.

Boyd, D. (2010). Social Network Sites as Networked Publics: Affordances, Dynamics, and Implications. In Z. Papacharissi (Ed.), *A Networked Self: Identity, Community, and Culture on Social Network Sites* (pp. 39–58). Routledge.

Bratteteig, T. (2010). A Matter of Digital Materiality. In I. Wagner, D. Stuedahl, & T. Bratteteig (Eds.), *Exploring Digital Design: Multi-Disciplinary Design Practices* (pp. 147–169). London: Springer.

Braw, E. (2023, May 24). Decoupling Is Already Happening—Under the Sea. *Foreign Policy*. https://foreignpolicy.com/2023/05/24/china-subsea-cables-internet-decoupling-biden/

Brink. (2000). Secular Icons: Looking at Photographs from Nazi Concentration Camps. *History and Memory*, *12*(1), 135.

Brock, J. (2023, March 24). U.S. and China Wage War Beneath the Waves—Over Internet Cables. *Reuters*. https://www.reuters.com/investigates/special-report/us-china-tech-cables/

Buchanan, B. (2020). *The Hacker and the State: Cyber Attacks and the New Normal of Geopolitics*. Harvard University Press.

Burgin, M. (2010). *Theory of Information: Fundamentality, Diversity and Unification*. World Scientific.

Burton, J., & Lain, C. (2020). Desecuritising Cybersecurity: Towards a Societal Approach. *Journal of Cyber Policy*, *5*(3), 449–470.

Buzan, B., Wæver, O., & Wilde, J. de. (1998). *Security: A New Framework for Analysis*. Lynne Rienner Publishers.

Chesnoy, J. (Ed.). (2015). *Undersea Fiber Communication Systems*. Academic Press.

Cybersecurity – Getting It Right. (2003). *Hearing of the subcommittee on Cybersecurity, Science, and Research, and Development, before the Select Committee on Homeland Security* (Serial No. 108-18), U.S. House of Representatives, 108th Cong.

Daskal, J. (2018). Microsoft Ireland, the CLOUD Act, and International Lawmaking 2.0. *Stanford Law Review*. https://www.stanfordlawreview.org/online/microsoft-ireland-cloud-act-international-lawmaking-2-0/

Deibert, R. (2015). The Geopolitics of Cyberspace After Snowden. *Current History*, *114*(768), 9–15.

Dembski, W.A. (2016). *Being as Communion: A Metaphysics of Information*. Routledge.

Dipert, R.R. (2010). The Ethics of Cyberwarfare. *Journal of Military Ethics*, *9*(4), 384–410.

Dourish, P. (2016). Rematerializing the Platform: Emulation and the Digital-Material. In S. Pink, E. Ardèvol, & D. Lanzeni (Eds.), *Digital Materialities: Design and Anthropology* (pp. 29–44). Bloomsbury Publishing.

Dourish, P. (2017). *The Stuff of Bits: An Essay on the Materialities of Information*. MIT Press.

Dunn Cavelty, M. (2008a). *Cyber-Security and Threat Politics: US Efforts to Secure the Information Age*. Routledge.

Dunn Cavelty, M. (2008b). Cyber-Terror—Looming Threat or Phantom Menace? The Framing of the US Cyber-Threat Debate. *Journal of Information Technology & Politics*, *4*(1), 19–36.

Eldred, M. (2013). *The Digital Cast of Being: Metaphysics, Mathematics, Cartesianism, Cybernetics, Capitalism, Communication*. Walter de Gruyter.

Euronews. (2024, March 14). *Which countries have banned TikTok and why?* https://www.euronews.com/embed/2499046

Evans, J., & Schneider, G. (2008). *New Perspectives on the Internet, Brief*. Cengage Learning.

Evens, A. (2015). *Logic of the Digital*. Bloomsbury Publishing.

Facilitating Cyber Threat Information Sharing and Partnering with the Private Sector to Protect Critical Infrastructure. (2013). *An Assessment of DHS Capabilities: Hearing before the Subcommittee on Cybersecurity, Infrastructure Protection, and Security Technologies, of the Committee on Homeland Security* (Serial No. 113-17), House of Representatives, 113th Cong.

Floridi, L. (2010). *Information: A Very Short Introduction*. OUP Oxford.

Floyd, R. (2011). Can Securitization Theory Be Used in Normative Analysis? Towards a Just Securitization Theory. *Security Dialogue*, *42*(4–5), 427–439.

Floyd, R. (2015). Just and Unjust Desecuritization. In T. Balzacq (Ed.), *Contesting Security: Strategies and Logics* (pp. 122–138). Routledge.

Fouad, N.S. (2024). Cyberbiosecurity in the New Normal: Cyberbio Risks, Pre-Emptive Security, and the Global Governance of Bioinformation. *European Journal of International Security*, 1–21.

Fradkov, A. (2007). *Cybernetical Physics: From Control of Chaos to Quantum Control.* Springer.

Fraser, E. (2016). Data Localisation and the Balkanisation of the Internet. *SCRIPTed*, *13*(3), 359–373.

Fresco, N., & Wolf, M.J. (2016). Information Processing and Instructional Information. In L. Floridi (Ed.), *The Routledge Handbook of Philosophy of Information* (pp. 77–89). Routledge.

Gellman, B., & Soltani, A. (2013, October 30). NSA infiltrates links to Yahoo, Google data centers worldwide, Snowden documents say. *Washington Post*. https://www.washingtonpost.com/world/national-security/nsa-infiltrates-links-to-yahoo-google-data-centers-worldwide-snowden-documents-say/2013/10/30/e51d661e-4166-11e3-8b74-d89d714ca4dd_story.html

H.R. 285: Department of Homeland Security Cybersecurity Enhancement Act of 2005 (2005). *Hearing before the Subcommittee Economic Security, Infrastructure Protection, and Cybersecurity, of the Committee on Homeland Security* (Serial No. 109-11), House of Representatives, 109th Cong.

Hallberg, B. (2009). *Networking: A Beginner's Guide* (Fifth Edition). McGraw Hill Professional.

Hansen, L. (2011). Theorizing the image for Security Studies: Visual securitization and the Muhammad Cartoon Crisis. *European Journal of International Relations*, *17*(1), 51–74.

Hansen, L., & Nissenbaum, H. (2009). Digital Disaster, Cyber Security, and the Copenhagen School. *International Studies Quarterly*, *53*(4), 1155–1175.

Harding, L. (2014). *The Snowden Files: The Inside Story of the World's Most Wanted Man.* Guardian Faber Publishing.

Harvey, R., & Weatherburn, J. (2018). *Preserving Digital Materials.* Rowman & Littlefield.

Hassib, B., & Alnemr, N. (2021). Securitizing Cyberspace in Egypt: The Dilemma of Cybersecurity and Democracy. In *Routledge Companion to Global Cyber-Security Strategy*. Routledge.

Heumann, S. (2017). Security in Cyberspace: The Limites of Nation-State Centric Approaches to Security in Global Networks. In J.D. Bindenagel, M. Herdegen, & K. Kaiser (Eds.), *International Security in the 21st Century: Germany's International Responsibility* (pp. 121–130). V&R unipress GmbH.

Hill, J.F., & Noyes, M. (2019). Rethinking Data, Geography, and Jurisdiction: A Common Framework for Harmonizing Global Data Flow Controls. In R. Ellis & V. Mohan (Eds.), *Rewired: Cybersecurity Governance* (pp. 195–230). John Wiley & Sons.

Huysmans, J. (2011). What's in an act? On security speech acts and little security nothings. *Security Dialogue*, *42*(4–5), 371–383.

Kallender, P., & Hughes, C.W. (2017). Japan's Emerging Trajectory as a 'Cyber Power': From Securitization to Militarization of Cyberspace. *Journal of Strategic Studies*, *40*(1–2), 118–145

Kallinikos, J. (2012). Form, Function, and Matter: Crossing the Border of Materiality. In P.M. Leonardi, B.A. Nardi, & J. Kallinikos (Eds.), *Materiality and Organizing: Social Interaction in a Technological World* (pp. 67–87). OUP Oxford.

Karnani, M., Pääkkönen, K., & Annila, A. (2009). The Physical Character of Information. *Proceedings of the Royal Society A: Mathematical, Physical and Engineering Sciences*, *465*(2107), 2155–2175.

Kitchin, R., & Dodge, M. (2011). *Code/space: Software and Everyday Life*. MIT Press.

Kittler, F. (1995). There is No Software. *Ctheory*, 10–18.

Krapp, P. (2011). *Noise Channels: Glitch and Error in Digital Culture*. University of Minnesota Press.

Lacy, M., & Prince, D. (2018). Securitization and the Global Politics of Cybersecurity. *Global Discourse*, 8(1), 100–115.

Landauer, R. (1991). Information is Physical. *Physics Today*, 44(5), 23–29.

Landauer, R. (1999). Information Is a Physical Entity. *Physica A: Statistical Mechanics and Its Applications*, 263(1–4), 63–67.

Lawson, S.T. (2019). *Cybersecurity Discourse in the United States: Cyber-Doom Rhetoric and Beyond*. Routledge.

Lloyd, S. (2000). Ultimate Physical Limits to Computation. *Nature*, 406(6799), 1047–1054. https://doi.org/10.1038/35023282

Lombardi, O., & López, C. (2017). Information, Communication, and Manipulability. In O. Lombardi, S. Fortin, F. Holik, & C. López (Eds.), *What is Quantum Information?* (pp. 53–76). Cambridge University Press.

Long, M.L. (2023, May 31). *Information Warfare in the Depths: An Analysis of Global Undersea Cable Networks*. U.S. Naval Institute. https://www.usni.org/magazines/proceedings/2023/may/information-warfare-depths-analysis-global-undersea-cable-networks

Lutz, E., & Ciliberto, S. (2015). From Maxwell's Demon to Landauer's Eraser. *Physics Today*, 68(9), 30–35.

Malaspina, C. (2018). *An Epistemology of Noise*. Bloomsbury Publishing.

Masur, P.K. (2018). *Situational Privacy and Self-Disclosure: Communication Processes in Online Environments*. Springer.

Mihalache, A. (2002). The Cyber Space-Time Continuum: Meaning and Metaphor. *The Information Society*, 18(4), 293–301.

Misra, S., & Goswami, S. (2017). *Network Routing: Fundamentals, Applications, and Emerging Technologies*. John Wiley & Sons.

Mitnick, K. (2019). *The Art of Invisibility: The World's Most Famous Hacker Teaches You How to Be Safe in the Age of Big Brother and Big Data*. Little, Brown.

Overview of the Cyber Problem. (2003). *A Nation Dependent and Dealing with Risk, Hearing of the Subcommittee on Cybersecurity, Science, and Research, and Development, before the Select Committee on Homeland Security* (Serial No. 108-13), U.S. House of Representatives, 108th Cong.

Piccinini, G., & Scarantino, A. (2016). Computation and Information. In L. Floridi (Ed.), *The Routledge Handbook of Philosophy of Information* (pp. 23–29). Routledge.

Protecting America From Cyber Attacks. (2015). *The Importance of Information Sharing: Hearing before the Committee on Homeland Security and Governmental Affairs* (Serial No. 114-412), U.S. Senate, 114th Congress.

Reisman, D. (2017, May 22). *Where Is Your Data, Really?: The Technical Case Against Data Localization*. Lawfare. https://www.lawfareblog.com/where-your-data-really-technical-case-against-data-localization

Reuters. (2020, September 18). Prosecutors open homicide case after cyber-attack on German hospital. *The Guardian*. https://www.theguardian.com/technology/2020/sep/18/prosecutors-open-homicide-case-after-cyber-attack-on-german-hospital

Reuvid, J. (2006). *The Secure Online Business Handbook: A Practical Guide to Risk Management and Business Continuity*. Kogan Page Publishers.

Rid, T. (2013). More Attacks, Less Violence. *Journal of Strategic Studies*, 36(1), 139–142.

Robertson, J., & Riley, M. (2018, October 4). The Big Hack: How China Used a Tiny Chip to Infiltrate U.S. Companies—Bloomberg. *Bloomberg Businessweek*. https://www.bloomberg.com/news/features/2018-10-04/the-big-hack-how-china-used-a-tiny-chip-to-infiltrate-america-s-top-companies

Rosenberg, M., & Nixon, R. (2017, September 13). Kaspersky Lab Antivirus Software Is Ordered Off U.S. Government Computers. *The New York Times*. https://www.nytimes.com/2017/09/13/us/politics/kaspersky-lab-antivirus-federal-government.html

Rushe, D. et al. (2013, October 31). Reports That Nsa Taps into Google and Yahoo Data Hubs Infuriate Tech Giants. *The Guardian*. http://www.theguardian.com/technology/2013/oct/30/google-reports-nsa-secretly-intercepts-data-links

Securing Critical Infrastructure in the Age of Stuxnet. (2010). *Hearing before the Committee on Homeland Security and Governmental Affairs* (S. Hrg. 111-103), U.S. Senate, 111th Cong.

Securing the Modern Electric Grid from Physical and Cyber Attacks. (2009). *Hearing before the Subcommittee on Emerging Threats, Cybersecurity, and Science and technology* (Serial No. 111-30), U.S. House of representatives, 111th Cong.

Selby, J. (2017). Data Localization Laws: Trade Barriers or Legitimate Responses to Cybersecurity Risks, or Both? *International Journal of Law and Information Technology*, 25(3), 213–232.

Starosielski, N. (2015a). Fixed Flow: Undersea Cables as Media Infrastructures. In L. Parks & N. Starosielski (Eds.), *Signal Traffic: Critical Studies of Media Infrastructures* (pp. 53–70). University of Illinois Press.

Starosielski, N. (2015b). *The Undersea Network*. Duke University Press.

Straube, T. (2017). Situating Data Infrastrcutures. In R. Kitchin, T.P. Lauriault, & G. McArdle (Eds.), *Data and the City*. Routledge.

The Department of Defense. (2011). *Department of Defense Strategy for Operating in Cyberspace*. United States Government. http://nsarchive.gwu.edu/NSAEBB/NSAEBB424/docs/Cyber-050.pdf

The White House. (2003). *The National Strategy to Secure Cyberspace*. United States Government. https://www.us-cert.gov/sites/default/files/publications/cyberspace_strategy.pdf

Timpson, C.G. (2013). *Quantum Information Theory and the Foundations of Quantum Mechanics*. OUP Oxford.

Tse, P. (2013). *The Neural Basis of Free Will: Criterial Causation*. MIT Press.

Vonderau, P., & Holt, J. (2015). 'Where the Internet Lives': Data Centers as Cloud Infrastructure. In L. Parks & N. Starosielski (Eds.), *Signal Traffic: Critical Studies of Media Infrastructures* (pp. 71–93). University of Illinois Press.

Vuori, J., & Saugmann, R. (Eds.). (2018). *Visual Security Studies: Sights and Spectacles of Insecurity and War*. Routledge.

Wæver, O. (2009). What Exactly Makes a Continuous Existential Threat Existential—And How Is It Discontinued? In O. Barak & G. Sheffer (Eds.), *Existential Threats and Civil-security Relations* (pp. 19–36). Rowman & Littlefield.

Zhu, J., & Knoespe. (2007). Continuous Materiality: Through a Hierarchy of Computational Codes. *Proceedings of the Seventh International Digital Arts and Culture Conference*, 188–198. http://eleven.fibreculturejournal.org/fcj-076-continuous-materiality-through-a-hierarchy-of-computational-codes/

7   Conclusion

Cybersecurity has been transforming the security agenda of nations and non-state actors around the world since the 1990s, though it has not had the same transformative impact on theories and conceptual frameworks of security in International Relations (IR). The policy-oriented nature of cybersecurity as a field of research posed a theoretical challenge to Security Studies, albeit one that has been approached mostly as a challenge of inclusion – i.e., how to include cybersecurity in existing theoretical and conceptual frameworks – rather than as a question of deeper transformation – i.e., how cybersecurity may have transformed the meaning and practices of security. Many theoretical approaches to the study of cybersecurity produced important insights regarding its discursive particularities; but stopped short of investigating the materiality of the field and how it can transform the meanings and logic(s) of security beyond those of existing security theories.

Against that background, the book advanced an alternative theorisation of cybersecurity that acknowledges its peculiarity and inherent multidisciplinarity, in a way that does not simply bind it to the existing logics of other security fields. The book is thus a study of the ontology and materiality of cybersecurity as such, not just of how human actors perceive it or discursively construct it. It does not assume that the ambiguity of cybersecurity can be overcome only by empirical analysis. Instead, it problematises the very being of the "cyber" as an essential constitutive force of its (in)security. This is cybersecurity as an infosphere rather than a discursively constructed security sector whose peculiarity is not reduced to its novelty as just one additional sector that we can test the assumptions of existing theories on. It is different because it challenges such assumptions and produces different security logics that should be theorised for differently. This is what the book does by exploring the informational ontology of cybersecurity and the implications it has on our theoretical understanding of security.

Combining theoretical approaches from the philosophy of information, information theory, cybernetics, software studies, new materialism, and risk studies, the book conceptualised cybersecurity as *entropic security* that is governed by the logics of *negentropy, emergence,* and *noise*. This conceptualisation is based on three core assumptions that were fleshed out across different

DOI: 10.4324/9781003454113-7

chapters: the field of cybersecurity has an informational ontology; information is a peculiar entity; and thus, it follows that cybersecurity is equally peculiar in relation to other existing security sectors and should be studied through a different theoretical framework.

By way of extension, the theoretical exploration of cybersecurity developed here is based on three main pillars. Firstly, it approaches information as a vital, active force in the co-production of the security of the "cyber" and of the development of its technologies. Secondly, it assumes that security in the infosphere entails a deeper theoretical analysis of the referent objects of security that is not exclusively tied to human actors and their interests. Beyond listing a group of referent objects of relevance to human life, the book investigated information as the ultimate referent object of cybersecurity and explored the different forms of materialities it possesses. Thirdly, security in the infosphere does not follow the fixed security logics assumed by many of the theoretical literature in security studies and cybersecurity. It challenges existing understandings of existentiality, exceptionality, and emergency measures and transcends the traditional binary divisions between security and risk.

The matter and materialities of cybersecurity

Investigating the informational ontology of cybersecurity in this book is ultimately a study of materiality. Three forms of materiality were analysed. Firstly, the book approached *materiality as intrinsic properties* of information as an overarching assumption. By this the book means the peculiar properties that are inherent to the existence of information, and not necessarily ascribed to it by discursive utterances. In Chapter 4, the book examined the intrinsic uncertainties and tendency towards disorder that characterise information systems and how they co-shape the essence of "security" in cybersecurity, through the logic of negentropy. The second form of materiality in the book is *materiality as agency of influence*. Again, this is an organising idea for the whole book in shifting the focus from human actors to information and analysing its role in co-constructing cybersecurity. Chapter 5 in particular focussed on the peculiar complexities, non-linearities, and contingencies of syntactic information (codes/software) and their influence in co-producing the logic of emergence. Finally, the book approached *materiality as physicality* in discussing the ontological status of information as compared to matter. Particular focus on this question of physicality was given in Chapter 5, by examining the complex (non-)physicality of information and its generative influence on the logic of noise and mundane cybersecurity. Though the analysis of these three forms of materiality may intersect with other security or informational fields, it remains cybersecurity-specific. Accordingly, those three chapters (4–6) started with general theoretical exploration of theories of information, followed by an application on digital informational systems, and ultimately establishing a connection between these theoretical arguments and the empirical analysis of cybersecurity in practice.

First, the book examined information as a complex entity that is ontologically linked to uncertainty. In part, the complexity of information is a product of its multiplicity and transformational capacity. In operation, information is capable of transforming itself, its environment, and other agents' perceptions of it. This transformational capacity produces complexity and also variety, hence the definition of information as "reflected variety." Most importantly, the complexity of information is a result of the indeterminacies associated with its operation. As shown in Chapter 4, entropy – defined as uncertainty or disorder – has been integral to the conceptualisation of information, as part of the development of information theory. Mostly, information has been defined through entropy: either as its inverse or its synonym. In communication contexts, the mathematical theory of information assumed that indeterminacy is the default state that information transmission seeks to minimise. Because information and communication systems interact with thermodynamic subsystems, the probability of noise is always high. As a result, the outcomes that an information system produces are best described as *emergent* rather than *resultant*.

This creates many uncertainties in the security management of digital information systems. It is impossible to know all the vulnerabilities that exist in a system beforehand. Many bugs appear in a later stage of software operation or are discovered only once they are exploited in a hostile cyber operation. Even patching those vulnerabilities is an unpredictable process, as it is difficult to know how the patch would react with the system before its actual application. Intrusion detection is also made difficult by the innate multiplicity of information and the difficulty of keeping a static image of a system with a large number of attack targets. Attribution and damage analysis are other processes marked by vast uncertainties, due to the use of botnets, proxies, onion routing, etc. Also, in the case when information about vulnerabilities is available, uncertainties persists due to the lack of technical knowledge about those systems and their complexity on the part of end users. Likewise, software manufacturers may be reluctant to issue a certain patch before adequately testing it, fearing that they may lead to more vulnerabilities and bugs once applied.

Second, the book presented complexity and contingencies as intrinsic to the ontology of information. Information is sometimes defined as *the difference that makes a difference*, to signify its power to achieve *order*, *change*, and *causation*. Some approaches even assume a cosmic fundamentality that is linked to information, by either explaining evolution in informational terms or regarding information as reality per se. As the book explained, information and cybernetics were important catalysts to post-humanism and new materialism, by viewing humans as information processing entities that can be comparable to intelligent machines. Generally, information systems are purposeful; they have to retain some level of intelligence to operate. They may exercise autonomy by choosing among various options and taking decisions with little to no intervention from the human operator. They have varying degrees of proactivity and reactivity to their surrounding environments. The more complex the informational agent is, the more of these properties it possesses. Despite being a human creation, codes/software can infringe human control and act in unpredictable ways. Once put into action, codes/software operate independently of humans, and through advanced algorithms they can take many decisions on behalf of the user without consulting them in the process. This does not just apply to normal users, but even to programmers. Many codes/software are becoming very complex entities, produced by a large number of programmers, making it difficult for a single expert to claim complete understanding of how they operate.

Hence, codes/software are engineered rather than designed, since many of their functionality and potential errors only appear once they start operating. Malwares are an important type of codes/software whose potential unpredictability is maximised by their ability to self-replicate and self-perpetuate. Their operation reflects the non-linearity of codes/software and their constant state of emergence. Even when carefully designed and executed, malwares may propagate beyond the aggressor's control, spread to un-targeted systems, or produce unintended consequences. They can perform multiple self-preservation techniques that complicate security, such as stealthing, polymorphism, or metamorphism. That is why, cyber incidents caused by malware can be considered a major challenge to ideas of control upon which cybernetics and computing technologies were based. Security is no longer a matter of user's control over the system as it once was in the advent of ICTs.

Thirdly, information is approached in the book as a simultaneously physical and non-physical entity. It can only exist through physical representation, but also does not strictly follow the laws of physics. It is fundamentally different from matter and energy, even though it utilises both in its operation. Physicality is evident in the infrastructure that allows digital informational systems to function, including the different forms of hardware, devices, cables, routers, hubs, etc. Yet, digital matter is still not ordinary matter, because it is enabled by bits, codes, and protocols. Its functionality depends on those intangible elements of syntactic information that specify what can and cannot be done. On the other side, codes/software are not totally abstract, non-physical entities either. Their interaction with the physical can be seen, for instance, in the constant trade-off processes in their operation and evolution to account for the physical limitations of storage, power, and connectivity. Even the textual languages used in programming are dependent on vernaculars linked to physical experiences.

Although digital information is divisible, mobile, and can exist in multiple places simultaneously, its transcendentality is still influenced by a wide range of geopolitical considerations due to its equally important physical existence. Every device or software we use is manufactured somewhere by a private company that is subject to the legal systems of a certain country. Every time someone sends a message online, the data packets of this message take different routes that may cross borders regardless of the receiver's geographical proximity to us. As explained in Chapter 4, packet-switching is a decentralised process that is often decided by routers, not humans. Additionally, when we use a search engine to search for information, we are establishing a communication with a data centre that has a physical location we know nothing about. We may own our data, but we have no control on where it is at any given time. This all create various security and privacy issues that human users are involuntarily entangled in. The security implications resulting from these geopolitical considerations have proven more salient than futuristic scenarios of apocalyptic cyber conflicts.

In short, thinking about cybersecurity informationally allows us to employ a rich and multidisciplinary body of literature that introduces important insights to the study of the matter and materialities of this field. Cybersecurity as an infosphere is distinguished by information as its subject matter and referent object. "Information is information," as argued by Wiener (Wiener, 1948, p. 132). It does not resemble ordinary matter and is fundamentally different from other non-human entities. If all security fields have informational elements, cybersecurity is entirely informational. The intrinsic indeterminacies, complexities, contingencies, and (non-)physicality of information are all co-constitutive forms of materialities and are generative of cyber (in)security. They co-produce peculiar logics of security that are not reduced to discursive constructions; ones that the notion of entropy directly captures.

Cybersecurity as entropic security

Security is "an essentially contested concept" (Buzan, 1991). Traditionally, four main elements in conceptualising security constituted the foundation of this contestation in security studies: the referent objects of security, threat sources, the security agenda, and the link between security, threats, and dangers. These four elements formed a dividing line in the debate between the so-called "traditionalists" and their military/statist conceptualisation of security on one side, and the "wideners-deepeners" on the other side. The widening attempts broadened the security agenda to include social, environmental, economic, and other security topics, and deepened the analysis of the referent objects of security beyond the state. Nevertheless, this is a debate that runs within an anthropocentric framework that does not consider the materialities of non-human objects. It is humans that are the primary subjects of security; it is their discursive utterances and speech acts that construct it; and it is qualities associated with their lives that are the main referent objects of security.

In departing from this anthropocentric understanding of security, the book did not engage in a quest for alternatives for mere theoretical purposes. Rather, the theoretical exploration presented in the book is one that is imperative to grasp the peculiar materialities of cybersecurity and the specificity of its informational ontology. Acknowledging the materiality of information is a must in a field where even experts admit varying degrees of uncontrollability and unpredictability in managing the security of the systems in question. In so doing, the book revisited the fixed logics of security that some security theories presented, such as the securitisation theory, and that CCS criticised. Security that is approached as existential and exceptional should be approached differently once the peculiar materialities of information are considered. Emergency measures and enmity as human choices based on human interests and desires should be challenged. Additionally, the meaning of existentiality should be unpacked instead of taken for granted, and its relationship with urgency and physicality should be problematised. The result challenges the whole essence of security in cybersecurity and its intricate relationship with risk.

The book therefore introduced entropic security as an information-theoretic, *non-binary* concept that is capable of illuminating important ontological aspects of cybersecurity, that may not be fully grasped through the separate analytical frameworks of security and risk. The book used three definitions of entropy in constructing a trilogy of security logics to capture the complexity of cybersecurity: entropy as uncertainties and disorders (the logic of negentropy), entropy as randomness (the logic of emergence), and entropy as disruption in communication channels (the logic of noise). Each of these three logics was linked to one characteristic of information. That is, the logic of negentropy is co-produced by the indeterminacies of information systems; the logic of emergence is co-produced by the complexities of codes/software; and the logic of noise is co-produced by the simultaneous physicality and non-physicality of information.

Entropic security, and its three concurrent logics, reformulates the concept of security to account for the materialities of information. The three logics of negentropy, emergence, and noise are essentially linked by their resistance of the idea of *human control* of security and the centrality of *human intentionality*. Entropic security allows us to theorise for the generative capacities of information in co-producing a peculiar meaning for "security" in cybersecurity that is not reduced to human subjectivity. Accordingly, entropic security represents an ontological argument about cybersecurity that goes beyond security and risk as modes of governance. As such, entropic security is an overarching conceptualisation that is not reduced to moments of exception or the existence of particular threats/risks. Further, entropic security is a *semantic deviation* from the essentially positive connotations of the term "security" that does not adequately represent the complexities of cybersecurity.

In a practical sense, through its definition as uncertainty and disorder, entropy as a security analogy enables us to understand the intrinsic link between the (in)security of all users of information systems across geographical boundaries. The multi-stakeholder nature of cybersecurity, the non-geographical interdependencies that characterise cyber threats, and the fact that the security of all actors is as strong as their weakest link are all factors that resemble entropy's additive nature. In addition, through the analogy of entropy, we can theoretically analyse the paradox of the direct correlation between increasing cyber insecurity on one hand and the growing investments in technological development in general and cybersecurity in particular on the other. Further, defined as randomness and non-linearity, the analogy of entropy can capture the problems of targeting in cyber operations; the challenge of attribution; the contextual/relational aspects of the subjects and objects of cybersecurity; and the dilemma of responsibility/liability. Through the analogy of entropy, cybersecurity can be freed from the confines of the friend-enemy logic that characterise other fields like the military security. Though enmity is still part of the cyber threat perception, it does not always have to be anthropocentric; the enemy can be the vulnerability and the malware: code/software. Importantly, cyber defence does not always need a pre-defined enemy or an attack; it can be exercised against the entropic force of increasing disorder and insecurity. Finally, the analogy of entropy as noise is capable of highlighting the importance of mundane cybersecurity and why the urgency of cyber threats should not be bound to understandings of military attacks, existentiality, or high-profile incidents.

As discussed in Chapter 4, the uncertainties and disorders in information systems co-produce an understanding of cybersecurity as a moving target. Here, cybersecurity is entropic because of its tendency towards more insecurity, which is analogous to physical entropy's arrow of time. As a result of this entropic nature, cybersecurity is measured by the relative improvement in the insecurities of the future compared to those of the present. Security thus becomes a process rather than an end goal; it is the quality of the measures implemented rather than the state of being free from threats. This is because absolute security is unattainable and contradictory to the very nature of information systems. Vulnerabilities can be considered the by-product of complexity, and thus can only be managed and reduced, not eliminated. Prevention in the infosphere is not about stopping one big, major attack or threat, because the cyber threat has a continuous nature. Accordingly, defence in cybersecurity is better conceptualised in terms of negentropy (negative entropy) that represents anti-entropic practices aiming at countering the entropic force of disorder and uncertainties. Negentropy as the essence of cybersecurity is based on risk prioritisation and risk acceptance: accepting that some attacks will happen anyway, and thus directing most security measures and capabilities to the higher risks. Instead of aiming at the elimination of threats, anti-entropic cyber defence aims at shifting the point of absolute cyber insecurity further away and defying its inherent inevitability – just like physical negentropy aims at shifting the point of heat death away.

Cybersecurity is also entropic given its emergent and non-linear nature. The book introduced the logic of emergence in Chapter 5 to counter the assumptions of an in-control human that is implicit in the logics of emergency. Given the complex, dynamic, and decentralised nature of self-organising information systems, their emergent behaviour often leads to complex, dynamic, and emergent security. This emergent security can be seen first in the construction of enmity in cybersecurity. Hostile intents and capabilities alone are not as a strong legitimiser in the infosphere as they maybe in other sectors, particularly military security. In the infosphere, capabilities are difficult to quantify or observe, and are mostly dependent on the existence of exploitable vulnerabilities in the target's system and on the target's level of cyber dependency. Establishing a strong link between a threat and a particular enemy is also made more difficult due to the uncertainties of attribution. Secondly, malwares have the ability to propagate to unintended targets, leading to unintended consequences.

In addition, the logic of noise also contributes to the conceptualisation of cybersecurity as entropic. Just like noise is a challenge of information but one that disrupts rather than destroy, many of the threats that form the core of cybersecurity lie in the realm of the mundane in contrast to the existential. In the majority of security research, existentiality is seen as intrinsic to security, as an essential legitimiser to urgency/immediacy, and as more than physical. In cybersecurity, however, existentiality is *not as intrinsic* to security, *not as essential* for legitimising urgency, and is mostly *reduced* to the physical. Here, noise as disruption or interference in signal transmission in information theory can be used analogically to understand how cyber threats are constructed as urgent and immanent, without being existential. The majority of cyber threats resemble noise as a problem of communication in the sense that they are often *disruptive* rather than *destructive*. Existentiality does exist in cybersecurity debates, but just as *another* discourse in which assertions about destructions are usually limited to the military and intelligence. This does not mean that the cyber threat is not sometimes hyped, because hyping threats does not necessarily entail existentiality. It is the physical elements of information that make it possible for existentiality to register in cybersecurity discourses, due to the familiarity of the physical to our understanding of threats. Existentiality in the infosphere is connected to the possible physical damages as a result of a cyber-attack; that is why most of such discourses are focussed on the security of CNIs.

However, it is the simultaneous non-physicality of information that makes it difficult for existentiality to dominate the cyber threat perception. The general invisibility of cyber insecurity; the absence of adequate imagery; the replicability and retrievability of digital data; its capacity to exist in multiple places simultaneously; and the universality of digital devices are all important factors that challenge existentiality perceptions. Non-physicality too engenders uncertainties about whether an intrusion has taken place, identifying its starting point, and estimating the scale of the resulting damages even if/when an intrusion is detected. Nevertheless, just like noise in information theory, disruptive cyber incidents are seen as the parasite of information technologies in cybersecurity. They are normalised as an intrinsic element of the infosphere, yet one that needs to be resisted and minimised. This less than existential logic of noise is capable of invoking urgency and immediacy without existentiality and exceptionality.

Hitherto, all such conclusions are only valid when three methodological considerations are acknowledged. The first is expanding the empirical analysis to include cybersecurity practices and policies, not just discursive utterances or speech acts. The second is applying a multi-actor approach that considers the role of both state and non-state actors. These two points are important, because an approach that only analyses speech-acts of state actors would necessarily reach different conclusions about the security logics mentioned above. Existentiality, enmity, and emergency measures register more in speech acts than practices, and mostly in government discourses – particularly in those of the intelligence and the military. Thirdly, the assumptions and arguments presented by this book would not be applicable to a study of cybersecurity that only focusses on high-profile, government-backed cyber-attacks. While not denying their significance, these attacks are far less in number and frequency than all the other forms of cyber threats that constitute cybersecurity. Again, this is related to an adoption of a multi-actor approach that does not lock cybersecurity within security perceptions of the military and intelligence agencies. When private actors are included, the everyday and mundane cyber threats that might not get as much media attention would appear as important as the highly publicised ones in understanding the nature of security in this realm. That is to say, an approach that only focusses on speech-acts, state actors, and big cyber incidents, would still be confined to the conventional assumptions about security logics, even if attempted to study the materialities of the field.

Contributions, limitations, and prospects for further research

This book contributes to the theoretical understanding of cybersecurity as a field that remains policy-oriented and under-theorised in IR and Security Studies. It does so by employing an interdisciplinary approach that brings new insights to the study of cybersecurity from information sciences that have direct links to the evolution of its technologies. This is an approach that attends to the inherent multidisciplinarity of this realm that cannot be grasped by resorting to theories of security alone. Additionally, it established a theoretical link between the "cyber" and the "informational" beyond the traditional distinctions between cybersecurity and information security. This is meant to overcome the ambiguities of the cyber terminology and account for the arguable novelty of this field by problematising its ontology instead of stopping at the stage of conceptualisations and definitions. The book also contributes to the study of the materialities of cybersecurity by transcending the confines of anthropocentrism and representationalism. This was done to challenge perceptions of human control in constructing the security of information systems that evolved in paths humans could not fully envision; that operate in ways they cannot fully predict; and that produce threats they are not able to completely manage.

In a broader sense, the book also speaks to CSS, and STS. It presented a theoretical framework that problematises the referent object, and contextualises security logic(s) in constructing security as a process of co-production. Although this theorisation was specific to cybersecurity, it can be used to inductively study other security fields. Furthermore, the book is a contribution to the dialogue between new materialism and IR in general, and Security Studies in particular. By emphasising information as different from the *matter* that new materialism theorised for, the book can be considered an attempt to further develop ideas on the peculiarity of non-human "things," particularly in security construction. Importantly, it demonstrated the need for investigating the specificity of the different types of the non-human *things* instead of dealing with them as one homogenous category.

Moreover, through the notion of entropic security, the book shows that what matters for understanding security is not just the mere identification of referent objects as part of a security discourse. Instead, we need to study the ontology of the non-human referent object as a co-constitutive force in constructing the meaning and essence of security at large. This argument also speaks to CSS that attempts to incorporate new materialism or post-humanism in analysing security and questions of agency. It highlights the need for a study of security that goes beyond the mere inclusion of the non-human towards a search for alternative new materialist, post-humanist, or non-anthropocentric conceptualisations of "security" that dismantles its humanist underpinnings.

However, this informational framework to the study of cybersecurity has been developed in relation to US policy, so more research is needed to explore its applicability in different cultural and political contexts. further research is also needed to examine the possible transformational capacity of information in other security fields. The core argument of the book is that information is peculiar, and its peculiarity engenders a peculiar cybersecurity due to its informational ontology. This is done to challenge the perception of cybersecurity as just another sector in which the dynamics of other security fields can be detected. It would thus be interesting to take this argument further to see how the transformative capacity of information can contribute to cybersecurity's potential transformative influences on the general meaning, practices, and logic(s) of security in other fields. It is a need for investigating how just as cybersecurity has broadened the security agendas of all state and non-state actors around the world, it may have also interacted transformatively across other security sectors. Further research is needed to examine not just how security is different in the infosphere, but how the infosphere also transforms our general understanding of security beyond cybersecurity. Future research could explore too whether entropic security and its logics could be applicable to other security sectors, particularly to ones that are considered relatively "new," like food security, health security, etc. As put by Bruce Schneier in explaining the technical challenge of cybersecurity – a statement that can be extended to our theoretical understanding of security in IR:

We have some tricks, and we know how to avoid some obvious problems, but we have no scientific theory of security. It's still a black art and, although we're learning all the time, we have a long way to go.

(*Overview of the Cyber Problem*, 2003, p. 11)

References

Buzan, B. (1991). *People, States and Fear: An Agenda for International Security in the Post-Cold War Era*. Harvester Wheatsheaf.

Overview of the Cyber Problem. (2003). *A Nation Dependent and Dealing with Risk: Hearing of the Subcommittee on Cybersecurity, Science, and Research, and Development, Before the Select Committee on Homeland Security* (Serial No. 108-13), U.S. House of Representatives, 108th Cong.

Wiener, N. (1948). *Cybernetics: Or, Control and Communication in the Animal and the Machine*. Wiley & Sons.

Index